Radiocarbon: Calibration and Prehistory

Curves smooth and sinusoidal
Do not affect the Goidel,
And the feelings of the Celt
Are rarely for the svelte.

Sensual southern Angles
Suesspect the facts he mangles;
His short-term fluctuations
Arouse their animations.

Shapeless wiggles leave them cold:
'Illogical and over-bold'.
But up here in our Celtic hame
Their polynomials do the same.

Radiocarbon: Calibration and Prehistory

edited by
Trevor Watkins
for the University Press

EDINBURGH

22 George Square, Edinburgh

ISBN 0 85224 282 4

Printed in Great Britain by
Lewis Reprints Ltd, Tonbridge

Contents

Contents

Introduction

T. WATKINS

The six papers in this collection represent the formal contributions to a one-day seminar held in Edinburgh University on Saturday 9 March 1974. Some eighty people participated, and in spite of its hasty organisation and size the response on the day and reaction after the event suggested that the papers should be published and made widely available. As it happens they form a fairly coherent set, and are printed in the order in which they were delivered at the seminar. Coherent they may be, comprehensible too one hopes, for the purpose of the seminar was to bring the subject of the dendrochronological correction of C-14 dates to discussion in the light of recent research before a large, lay audience. But to comprehensiveness there is no pretension: in the compass of the seminar there was no time for more than six formal contributions, and the contributors were those who were able to accept the invitations. Undeniably a bias thus appeared, and as a result it would seem that the seminar was against those who smooth their curves, such as Colin Renfrew and Martin Clark, and for wiggly curves—perhaps not as rich in 'Schwung' as that of Suess, but wiggly none the less. The feeling at the seminar was also against premature attempts at definitive correction of C-14 dates.

Since the inception of the C-14 dating method there have been difficulties; some have not believed in the method because of the early dates indicated, and others who have accepted the method have misused the dates. Having assumed for long enough that C-14 dates were absolute dates, that dream of all archaeologists, we had to learn that the results of C-14 assay were mere expressions of probability. And then we were told that a very real possibility existed that 'C-14 years' were not necessarily the same as calendar years, and that the difference between the two was probably not simple and constant, but variable. We learned that C-14 dates as given by the laboratories were likely to be low (usually low rather than high), and that the real age range of samples could be older by a number of centuries, nearly a millennium perhaps. Most of us are more sophisticated than we used to be in our understanding of raw C-14 dates, but how are we to find out about C-14 correction? Since many people now claim to be using corrected dates in print, and since no single, agreed and accepted means of correction exists, it is not surprising that the situation seems confused. To most of us the specialist conferences which have taken place under the auspices of the Royal Society and British Academy in London, the Nobel Symposium in Uppsäla and at Lower Hutt in New Zealand are unhelpful, and our seminar in Edinburgh was primarily intended to allow discussion of the subject of correction before a lay audience in terms which they could understand.

The starting point of the first paper is the need for seeking a correction of C-14 estimations of age, and the current state of the art as seen by Richard Burleigh, who has been deeply concerned at the British Museum Research Laboratory with attempts to match expected against observed in terms of C-14 dates. His message will not be unfamiliar to those who have followed his contributions to *Antiquity* and the *Journal of Archaeological Science* of late: it is too early to use the curves at present being bandied around as correction curves for archaeological dates. It is bad enough having two half-lives for C-14 without adding a whole new range of possibilities introduced by the various curves and tables.

John Fletcher, based at the Oxford Research Laboratory for Archaeology and the History of Art and known for his work on mediaeval dendrochronology, puts the now famous bristlecone pine into its proper perspective, and suggests how we might better serve our archaeological interests through the combination of C-14 and dendrochronological dating techniques. His optimistic assessment of the prospects of obtaining long European sequences of absolutely dated tree-rings was a surprise to most archaeologists at the seminar, and hinted at the possibility of more direct tree-ring checks on Old World C-14 dates.

From original doubts about Suess's short-term fluctuations Barbara and Jim Ottaway sought a means of discovering an independent check on the 'smoothness' or 'wigglyness' of the C-14 correction curve. Their answer is that, short-term fluctuations apart, there are wiggles in the curve; their means of searching for regularities and irregularities in the curve is both novel and essentially simple, and incidentally provides a check independent of dendrochronology.

Historical chronologies in certain parts of the Old World provide yet another check on C-14 chronology, as has been well known from the outset. Both Anthony Snodgrass and Hugh McKerrell are concerned about the divergence between 'historical' dates and 'corrected' C-14 dates in the second millennium BC. Dr Snodgrass sets out to show that there is a body of C-14 dates from the Mycenaean period, in the light of which any attempts to compare 'corrected' C-14 dates for hither Europe with 'historical' dates for the

Aegean world are unwarrantable exercises in not comparing like with like.

For most of the historical C-14 dating-range, however, we are denied the luxury of checks and must rely solely on C-14 dated tree-rings of known age. But how to make sure of all the dendrochronological data? The answer is, by statistics. Hugh McKerrell's essay carefully takes us through the problems and the statistical solutions proposed. For the first time the data is set out, and the comparison of the various techniques is at last possible without having to have a battery of learned volumes to hand. Dr McKerrell concludes that the dendrochronologically corrected C-14 dates and the 'historical' dates are incompatible over the third and second millennia BC, and prefers the correction based on Egypt's astronomically derived historical chronology.

If as editor I may enter the lists, I should here endorse part of Hugh McKerrell's argument. The widely accepted Mesopotamian 'historical' chronology, which is derived independently of the Egyptian, should and does correspond to the Egyptian chronology at a number of independent but interlinked points. Thus, if one prefers the C-14 calendar calibrated by bristlecone pine tree-rings, one must not only reconcile the Egyptian calendar but also the Mesopotamian by the same amount. Unfortunately there are rather few Mesopotamian C-14 dates, but in general their trend is similar to those from Egypt. It is difficult to believe that the Egyptian chronological system is so much in error, but to be asked to accept that the Mesopotamian system, whose ultimate derivation is also astronomical, is wrong by exactly the same amount stretches my credibility beyond bursting point.

Andrew Fleming's contribution speaks for itself and, fittingly for a final contribution, leads the discussion out of the perhaps myopic view of tree-rings to a broader view of the wood as a whole. It is a salutary reminder that while chronology will always be important, we are no longer so consumingly concerned with it as archaeologists. Hopefully, we know that precedence in time is not in itself explanatory of cause and effect: to know that agriculture was practised here earlier than there is not to explain its spread, or even to suggest that it did spread. With chronology in its right place archaeologists can devote themselves to the real issues.

At the end of the book are three appendixes, which will be of use to different archaeologists. Hugh McKerrell has compiled sets of tables comparing the various current correction curves including the MASCA curve, which seems to be particularly widely favoured. By means of these tables one can compare the different curves, and obtain preliminary ideas of the sort of corrections that will apply. The second appendix has been contributed by Douglas Harkness, Director of the Radiocarbon Laboratory of the Scottish Universities Nuclear Reactor Centre at East Kilbride. It recommends how samples for C-14 dating should and should not be taken, recorded, treated and transmitted, so that archaeologists, who expect first-class results from the laboratories, can assist by submitting first-class material where possible and can treat the dates decently when received. Finally there is a select list of radiocarbon laboratories currently active on archaeological work.

Calibration of C-14 Dates: Some remaining uncertainties and limitations

R. BURLEIGH

This article, which is an abbreviated version of a talk given at the Edinburgh seminar, attempts to review the problem of the interpretation of radiocarbon dates in terms of calendar years. A brief outline is given of the factors that make such correlations necessary and of the work that has so far been done to make them possible. The calibration of the C-14 timescale very largely depends at present on the bristlecone pine chronology, which has already revolutionised the interpretation of many patterns of C-14 dates and in general undoubtedly provides the prehistorian with more acceptable chronologies. At the same time it is clear that many detailed uncertainties still remain and, accepting overall the new picture that has emerged, two principle reservations are still necessary. These are, first, that for the present only rather tentative calibrations of individual C-14 dates should be made and, second, that there is no immediate prospect of a definitive calibration relationship between C-14 and calendar years being established to enable more exact corrections to be made. The reasons for these two statements will now be discussed more fully. Obviously space does not permit an exhaustive analysis of all the factors which may be involved, and the more specialised literature of C-14 dating must be consulted to obtain a better idea of these. A number of references to such sources are indicated.

The physical principles on which C-14 dating is based are well attested and need not be restated in detail here. However, a large body of information has now been accumulated which shows that one of the fundamental assumptions of the method is no longer tenable in its original form. The idea that the rate of production of natural C-14 in the earth's atmosphere has always been essentially constant is now known to be untrue, so that C-14 years are not necessarily always equivalent to calendar years. This realisation has come about partly as a result of improvements in the accuracy with which the physical measurements can be made, but mainly because of observed discrepancies between the C-14 and the true dates when materials of known ages are measured. At present, independent check measurements extend back to about 5400 BC. Very roughly, there are relatively small differences for most practical purposes between C-14 and calendar ages from the present until about the middle of the second millennium BC. Earlier than that, C-14 dates are found to be increasingly too young, up to a maximum of about 800 years in comparison with the true age. There is also evidence of shorter-term fluctuations superimposed on the long-term trend. Detailed knowledge of these variations has come from systematic measurement of the natural C-14 content of dendrochronological material, in particular the very long-lived bristlecone pine which grows in parts of the SW United States. Independent general confirmation of the occurrence of variations has also been provided by C-14 measurements of historical material from closely dated contexts, notably from Egypt, back to c. 3100 BC.

The causes of natural C-14 variations are not yet understood with complete certainty, but it seems probable that the long-term changes are closely linked with known gradual changes in the intensity of the earth's magnetic field. These will have affected the amount of cosmic radiation reaching the earth's atmosphere from outer space, leading in turn to changes in the rate of production of C-14. Changes in solar activity may account for some of the shorter-term variations and there may have been internal changes in the behaviour of the parts of the carbon exchange reservoir, i.e. the oceans, the atmosphere itself, the biomass, etc., linked with climatic change. Some or all of these variables may have interacted in complex ways at different times. The variations have become the subject of much intensive study over the last decade or so, not only to establish a correlation between C-14 and calendar years but also because they are of great geophysical interest. The Proceedings of the 12th Nobel Symposium in 1969 (Olsson 1970) and of the more recent 8th International Conference on Radiocarbon Dating (Rafter and Grant-Taylor 1973) are probably the most comprehensive sources to the published literature at present. Much of this is, however, of a technical nature and does not deal primarily with the kinds of implications which are currently under discussion in many archaeological journals and publications.

Nearly all of this work has been based on data obtained from C-14 measurement of bristlecone pine wood. The bristlecone pine grows in a number of states of the American south-west, but interest has centred mostly on trees growing under climatic stress above 3000 metres in a limited area of the White Mountains of east-central California. Survival of the oldest trees appears to have been favoured there, and some living individuals have been found to exceed 4500 years of age. Originally the bristlecones were of interest mainly from the point of view of the study of climatic change as reflected by the varying width of the annual rings. But, because of the exceptionally complete record which this material holds of the past changes in atmospheric C-14 levels, the emphasis understandably shifted towards cross-checking of the C-14 timescale. The dendrochronology of the bristlecone pine extends back to about 8200 years before the present and is claimed to be accurate to within one year (additional dendrochronological work now appears to have substantiated this claim: see *Nature 248*, no. 5444, 8 March 1974, p. 104). Wood from the major part of the earliest millennium is too valuable to sacrifice for C-14 measurement yet and this must wait until more material in that age range has been identified.

As a result of progressive reporting of their work by the principal dating laboratories concerned up till now with this problem (Arizona, La Jolla, Pennsylvania) various calibration curves or tables have been published (e.g. those of Suess 1970; Damon *et al.* 1972; Michael and Ralph 1972). These laboratories have also published more provisional versions of their findings, and some other more generalised attempts have been made independently to construct calibration relationships from the published data. Very generally, most of these calibrations differ from one another mainly in detail, and all agree as to the main trend of C-14 deviation shown by the tree-ring material. More recently Ralph *et al.* (1973) have published a comprehensive calibration, which incorporates virtually all the tree-ring measurements made by the three principal laboratories, for the period from 1849 AD to 5383 BC. This probably constitutes the most useful generalised calibration so far available for present purposes. However, aside from the possibility of systematic differences between laboratories, there are a number of other uncertainties yet to be resolved. For example, there remains the question of whether the bristlecone calibration itself is generally applicable. As mentioned earlier it already seems clear, if not perhaps always completely accepted, that, at least for prehistoric Europe, dates corrected from the generalised bristlecone calibration provide more plausible interrelationships than do some of the conventional archaeological chronologies employed hitherto (Renfrew 1973a). Nevertheless, until it is certain that this calibration is free, for practical purposes, from any effects arising from the rather special nature of the bristlecone growth conditions (Harkness and Burleigh 1974) these new chronologies must be regarded as provisional (as indeed their chief proponents have stated). It seems unlikely on present evidence that these new frameworks will be in error by more than about 200 years at most, although there is also some evidence to suggest that a tendency for over-correction may result from the use of the bristlecone calibration (McKerrell 1972). Finally, there is the possibility that a limit of resolution may be imposed on C-14 dating by the suspected short-term C-14 fluctuations. There are no good reasons for supposing these may not have occurred, or conversely that, for example, Suess' curve shows them correctly, or that some other curves may not be too smoothed (Baxter 1974).

Some of these problems may eventually be at least partly resolved by means of the remarkably well preserved fossil oak and pine wood recently recovered from Irish peat mires (Smith *et al.* 1972). C-14 measurements have indicated that this material spans the period from about the beginning of the Christian era to around 7000 years bp, and there are indications that it may yet be possible to extend both ends of the sequence. When the dendrochronology of this material has been worked out in precise detail an exceptional, possibly unique, chance will be provided to cross-check the bristlecone data by a similar series of C-14 measurements at closely spaced intervals. Most importantly this comparison will be based on materials from two widely separated geographic regions, the latitude, altitude, climatic and growing conditions of which are completely different, apart from important differences between the trees themselves.

In the long term it will be desirable to calibrate C-14 dates accurately in terms of calendar years, and presumably this will eventually be done routinely by laboratories for all new dates. It remains to be seen exactly what limitations these calibrations will finally have. Corrections cannot be made in any other than a tentative way until there is agreement among laboratories, particularly those actually concerned in the systematic study of C-14 variations, on a final calibration relationship that also takes into account the latest half-life value. In the meantime preliminary calibrations are certainly necessary in order to arrive at greater absolute accuracy (although the *precision* of such dates will be lower than that of the 'raw' uncalibrated dates). Preliminary calibrations may also be needed to assess whether the raw dates fall into a section of the curve which may be more or less sensitive due to the occurrence of short-term fluctuations or 'kinks'. Otherwise, for the time being and for most ordinary purposes, C-14 dates should strictly be used and compared together directly in terms of C-14 years bp based on the 5570 year half-life as recommended by the journal *Radiocarbon* (for consistency using the lower case notation for radiocarbon years proposed by *Antiquity 46*, no. 184, 1972, p. 265). At present one cannot go very much further than this, whatever manipulations are performed on the bristlecone data alone.

European Dendrochronology and C-14 Dating of Timber

J. M. FLETCHER

Tree-ring chronology was being used extensively to date archaeological timber several years before the principle of C-14 dating was enunciated by Libby. Indeed, during a tour of Arizona and New Mexico in the summer of 1935 Libby and I visited cliff-dwellings for which dates had been obtained by Douglas from tree-ring studies on the beams (of ponderosa pine) that had been used in their construction.

Though there has been relatively little work in Britain on dendrochronology until recently, this science was firmly established on the continent of Europe in the 'pre-C-14' era. There were tree-ring studies on conifers of current growth in various parts of Scandinavia in the first decades of this century, but European dendrochronology was evolved by Huber (1941), who started his research on the widths of annual rings in 1938 at the Forestry-Botany Institute near Dresden and from 1946 onwards continued it at Munich. By the 1960s there were similar studies by Kolchin (1962) in Russia and at several other centres in western Europe (figure 1): for example by Hollstein (1965) at Trier; by Munaut (1966) at Louvain; and by Bauch, Liese and Eckstein (1967) at Hamburg. The trees studied have included both deciduous species (such as oak, ash, beech, alder) and coniferous ones (silver fir, larch, spruce, pine) that flourish in many areas of Europe. A summary of the research up to 1972 has been published (Eckstein 1974).

From the German laboratories in particular have come impressive contributions to architectural and art history as well as to the dating of archaeological materials and the study of geomorphology in the post-glacial period. The provenance of those samples related to the works of man include areas as far apart as Switzerland in the south to Schleswig-Holstein in the north; as France (and even England) in the west to Slovakia in the east. In age they embrace not only Roman but Bronze Age and Late Neolithic material.

At the International Congress on Quaternary Geology in Zurich in 1956,

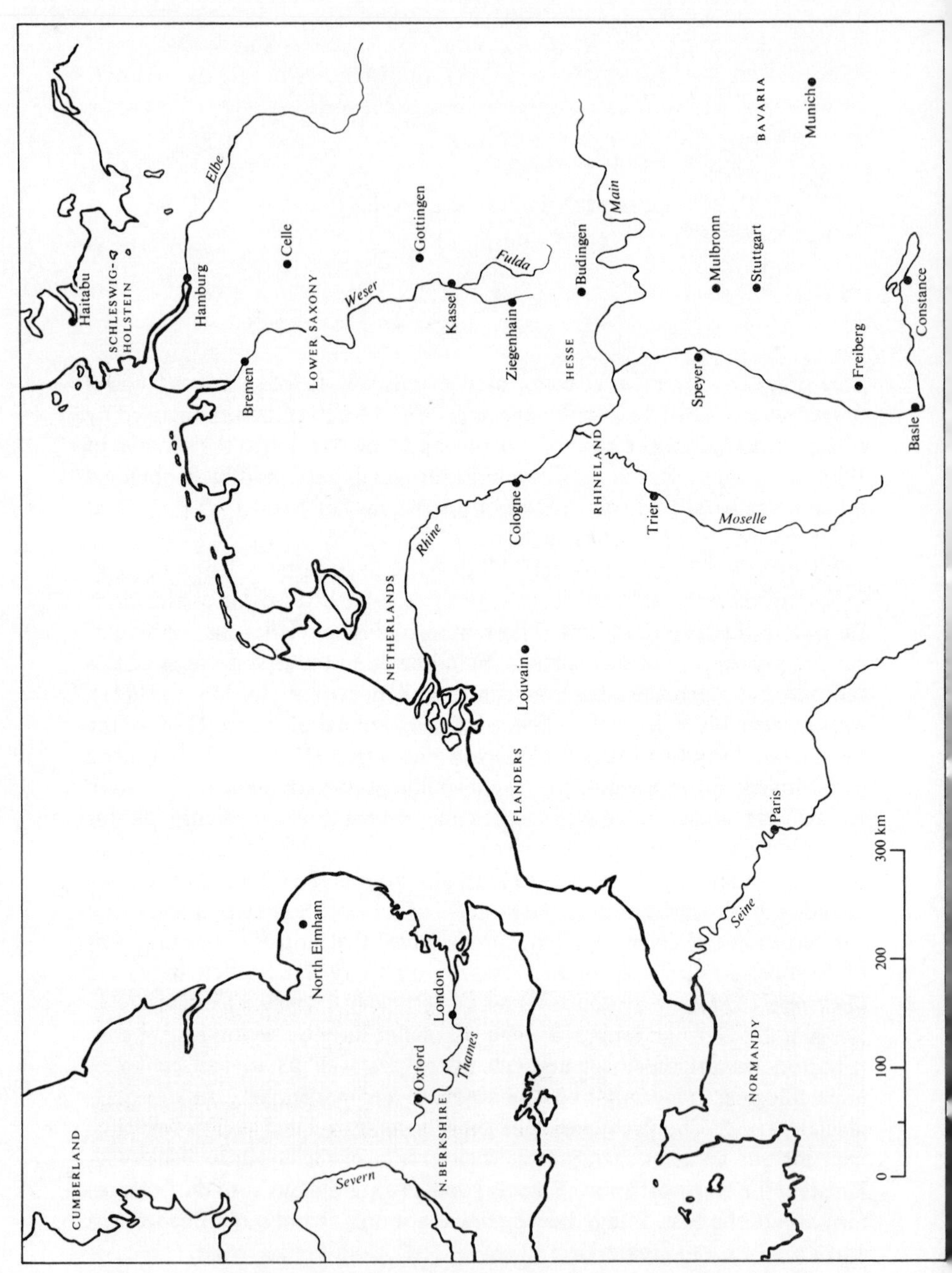

Figure 1. Western Europe showing places and areas mentioned in the text.

when C-14 dating was the sensational new technique, Huber was asked by a colleague whether he would abandon his work on dendrochronology (Huber 1970, 204). He gave two reasons, still valid, for not doing so. First, the greater accuracy (at least one hundred times, he then said) of the scale of a tree-ring chronology; secondly the need for it, as an independent method, to prove (or disprove) the constancy of the C-14 content of the atmosphere in times past, this being the factor on which the accuracy of the C-14 clock depends.

It is convenient to present this paper as an elaboration of the points made by Huber. First, examples are given of the relative accuracy of dates derived from the tree-ring reference curves that exist for central and northern Europe. Secondly, attention is drawn to the important role that European samples of tree-ring dated wood are playing in determining fluctuations in the level of C-14 in the atmosphere. Thirdly, the reasons for the inaccuracy and uncertainty of the conventional C-14 ages for wood are given, and the simplicity of correcting those of the last 2000 years to calendar dates is illustrated.

Dating by Dendrochronology in Central and North-Western Europe

Some European species, such as alder, are unsuitable for dendrochronology because of missing or false rings, others because of the absence of trees of adequate age. Fortunately these limitations do not apply to some of the widely-available species such as sessile and pedunculate oaks (*Quercus petraea* and *robur*), silver fir (*Abies alba*) and beech (*Fagus silvatica*). Fluctuations in the widths of the annual growth rings (plate 1) form the basis of the method widely adopted in Europe. At the end of his career Huber (1970) summarised his experience of this method gained over a period of thirty years, and found it appropriate to quote fifty-six references of German origin. A briefer version (Huber and Giertz 1970) was presented to the Conference on Scientific Methods in Medieval Archaeology held at Los Angeles in 1967.

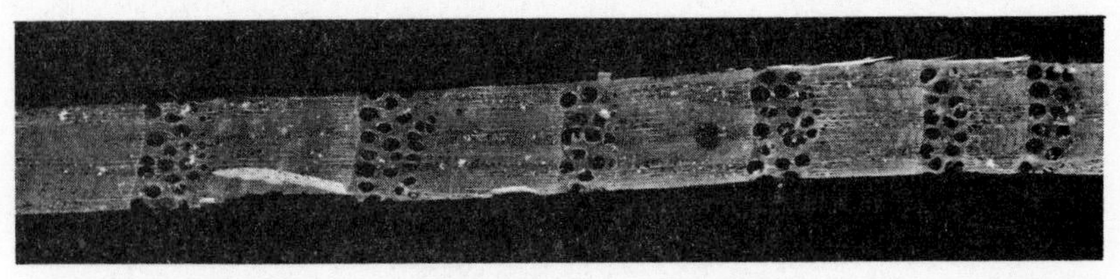

Plate 1. Five complete growth rings of oak, a ring-porous species. Enlargement of thin section ×10. Note the *line* which marks the beginning of each year's growth; the early wood (open vessels); and the late wood (no large vessels).

The first target is to produce a mean curve in which the year-to-year fluctuations in the annual widths of several trees are averaged. This forms a master or reference curve. It usually includes groups of years, each known as a 'signature', which contain some prominent feature. Several signatures occur between 1390–1430 AD (figure 2) in the thousand-year-long sequence derived for sessile oaks that grew in central Germany (Huber and Giertz 1969). Similar signatures occur in the reference curve for oaks that grew in SE England and Flanders in the same period (Fletcher, Tapper and Walker 1974).

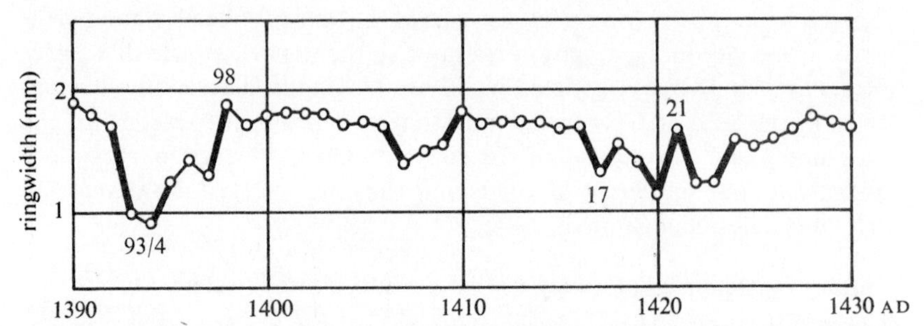

Figure 2. Reference curve for mid and southern Germany for oak with signatures in the period 1390–1430 AD (after Huber and Giertz 1967).

Dating by dendrochronology is achieved if the fluctuations on the ring-width chart for the sample of unknown date can be fitted with those of a reference curve for the same species and the same region. To find if there is such a matching position (figure 3) both visual comparison (using the signatures) and computer comparison (using the direction and sometimes also the magnitude of the changes in the two curves) are employed.

Other characteristics which relate growth to particular years can be helpful. One is wood density, determinable by X-radiography (Polge and

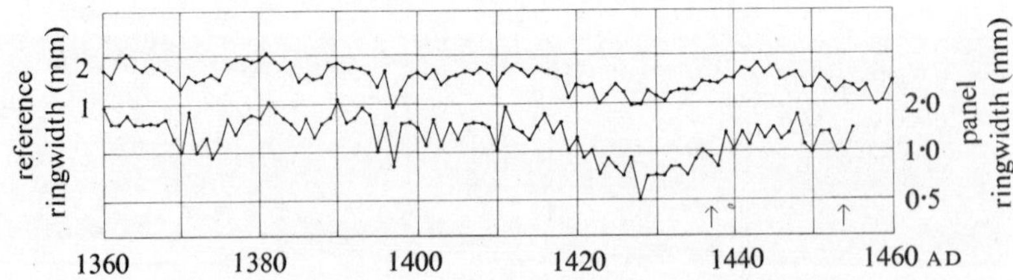

Figure 3. Below, the latter part of the chart for the ring-widths of a panel (used for portrait of Elizabeth Woodville in the Royal Collection at Windsor Castle). It is placed where it matches a reference curve (upper chart) based on 18 Anglo-Flemish trees of average rate of growth.

Keller 1969; Fletcher and Hughes 1970). Another (plate 2) shown by oaks that grew relatively slowly in SE England and Flanders, is the abnormal smallness of the early wood vessels in particular years in the second quarter of the 5th century (Fletcher 1975).

The master curve for the Spessart oaks derived by Huber and Giertz (1969) for the area around Hesse (figure 1) applies over an area of some 300 km square, while for less elevated and maritime parts of Europe there are other reference curves, some of which are given in table 1, that apply to

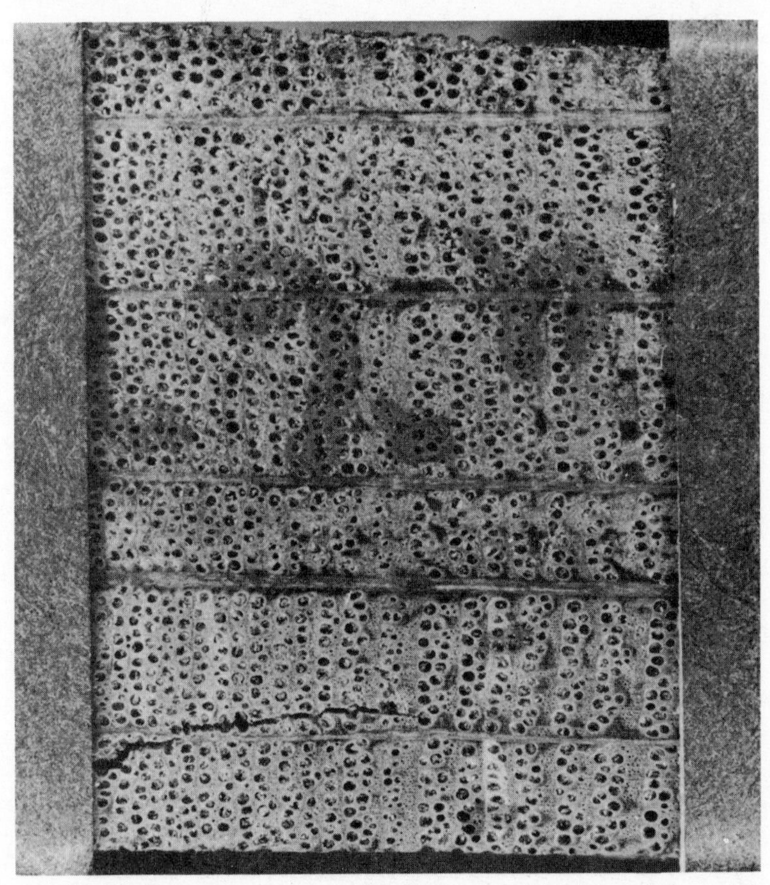

Plate 2. Part of the edge of a quarter-sawn oak board (growth from left to right) showing (i) on the left half, 15 successive narrow rings each with early wood wider than late wood; (ii) in the centre, a year (1437) in which the early vessels were abnormally small; (iii) lengthwise in the radial direction medullary rays, five in number; these contribute to the reserves of energy.

Table 1. Dendrochronology: Some reference curves of ring widths for central and north-western Europe. For a more complete list covering the whole of Europe see Eckstein (1974).

Species	Region	Years AD	Samples* from	Reference & Origin
oak	Germany, mid & southern	960-1960	churches, castles, houses	Huber & Giertz 1969, Munich
oak	Germany, west of the Rhine	822-1964	churches, weirs, mills	Hollstein 1965, Trier
oak	Germany, Schleswig-Holstein	1352-1969	churches, bell-towers, barns	Eckstein, Bauch & Liese 1970, Hamburg
oak	Germany, southern Weser	1004-1970	churches, castles, houses, bridges	Delorme 1973, Gottingen
oak	SE England and Flanders	1320-1550	churches, panel paintings, chests	Fletcher, Tapper & Walker 1974, Oxford
larch	Bavaria and Austria	1340-1947		Brehme 1951, Munich
silver fir	central Europe	820-1969	churches, bridges, houses	Becker & Giertz 1970, Hornstein 1964/5, Müller-Stoll 1951, Munich

* Other than from modern trees. The southern Weser curve, for example, was derived from 86000 ring widths on 556 samples.

smaller areas. Indeed, as that table indicates, the process of constructing a series of reference curves for use in almost all parts of central and north-western Europe where oaks have flourished and have been employed by man is well under way.

While much attention has rightly been given at conferences and in publications to dating by C-14 and to the contribution made by samples of wood from the bristlecone pine to its calibration, the numerous achievements in Europe of dating by dendrochronology have tended to be overlooked, at least in Britain. Some examples in the fields of archaeology and of architectural and art history are given in table 2. In all they amount to many hundreds, and in providing meaningful dates for sequences of occupation or cultures must compare in number, as well as exceeding in accuracy, those obtained in the same regions by C-14.

The information provided by tree-ring dating is the year (or likely period) in which the tree was felled and hence, indirectly, the date in which the timber from it was used. The large timbers found in ancient buildings, wells, piles for bridges, weirs and mills were almost invariably used in Roman and subsequent times within a year of being felled (Hollstein 1965). Hence if the bark remains, a construction date accurate to a year can be derived. Neiss (1968) achieved this for several phases in the construction of the castle and church at Büdinger from studies on their oak members.

Often, however, neither the bark nor all the sapwood remains with the timber. In oak, sapwood (plate 3) is attacked by insects and tends to disintegrate in air, while heartwood is unaffected. For certain uses most or all of the sapwood was therefore removed. It occupies about an inch in width and may cover as few as eight years (if the rings are very wide) and as many as 35 (if very narrow). For oaks of average width and about 200 years old, a value of 20 ± 6 years has been found. Hence if the heartwood-sapwood interface exists, the felling year can be estimated to within a decade or so. In the following paragraphs examples with and without surviving sapwood are given.

The Bremen cog. In 1962, in the course of dredging the River Weser at Bremen to form a new dock, this sea-faring trading vessel (Fliedner and Pohl-Weber 1964) was discovered buried in the mud. The ship (plate 4), some 78 feet long, was still being fitted out when it broke away from its moorings and sank. The annual rings of one of the oak members (Trunk II) were found by Bauch (1969) to include not only twenty years of sapwood but the interface between the sapwood and bark. The ring-widths were matched and dated by the reference curve for central and southern Germany, the similarity being greatest with its component for the hilly area around Kassel, which lies some 200 km south of Bremen. In this way it was found that the felling year of the tree was 1378 and its likely origin

Table 2. Some applications of European dendrochronology.

	Location	Species	Reference
A. *In Archaeology*			
1. Neolithic lake settlements	Switzerland	oak	Huber & Merz 1963
2. Bronze Age settlements	Switzerland	oak	Huber & Merz 1963
3. Celtic & Roman bridges & fortifications	Rhineland	oak	Hollstein & Cüppers 1967
4. Medieval houses & pavements	Novgorod	pine & spruce	Kolchin 1962
5. Viking settlement	Haithabu, N. Schleswig	oak	Eckstein & Liese 1971
B. *In Architectural History*			
1. Medieval buildings	Budingen, Ziegenhain, etc., in Hesse	oak	Neiss 1968
2. Historic buildings	Schleswig-Holstein	oak	Eckstein, Bauch & Liese 1970
3. Medieval cathedrals	Trier, Speyer	oak	Hollstein 1968
4. Medieval cathedrals & abbeys	Constance, Freiburg & Maulbronn	fir	Hornstein 1964/5
5. Medieval churches	Landshut, Bavaria	fir	Becker 1968
6. Post-medieval buildings	Norway	pine & spruce	Slastad 1957
C. *In Art History*			
1. Panel paintings	Netherlands, of 16th, 17th & 18th centuries	oak	Bauch, Eckstein & Meier-Siem 1972
2. Panel paintings	SE England & Flanders, of late 15th & 16th centuries	oak	Fletcher 1974

Plate 3. Polished surface of part of oak beam showing (i) growth rings, each of one light and one dark band; (ii) their variable width; (iii) to left, heartwood, dark in colour; (iv) to right, sapwood, lighter in colour and with holes due to insect attack. Bark is absent, but it would have been at extreme right.

Plate 4. The Bremen Cog: part of the port side after partial clearance. (Focke-Museum Bremen.)

one of the forests to the south of Kassel, from which logs could be floated down the Fulda and the Weser to Bremen. With an allowance of a year for the river transport and with microbiological evidence from the wood indicating no prolonged storage, the construction date could be placed very close to the year 1380.

Panel paintings. The second type of accuracy is illustrated by the oak panels which were used as supports for paintings. The likely period at which a panel was used can often be assessed to within a decade or so by dendrochronology. With most panels, all the sapwood was removed in making the board, but the craftsmen sometimes went so close to the heartwood/sapwood edge that in places some sapwood was left. Centuries later it is readily detected by the insect attack that has occurred.

On the portrait of Thomas, 2nd Lord Wentworth (National Portrait Gallery no. 1852) painted in 1568, several rings of sapwood were, exceptionally, left on both of its boards. Tree-ring analysis showed that the sapwood/heartwood boundary on the wide board (with 193 rings) occurred at 1546–1547. The average width of the rings here and in the sapwood was 1·3 mm, a value which implies that there were about twenty sapwood rings and that the likely date of felling of the tree was about 1566. The interval prior to the date of the painting was therefore only a very few years, in line with what has been found for other paintings (Bauch and Eckstein 1970). This information has been applied to early paintings of unknown date. For example, in the Royal Collection at Windsor Castle and mentioned in the first inventory, made in 1542, there are unsigned and undated portraits on oak panel of Henry V, Henry VI and Richard III that have been the prototype for many subsequent copies (Millar 1963). They are of similar size and presumably by the same artist. Each panel consists of two boards, one being quite narrow. We found (Fletcher 1974) the ring-width pattern for the three wider boards to be so similar to one another that they must have formed part of the same tree. The three charts were dated as follows:

for the Henry V board, 236 rings spanning the years 1259–1494;
for the Henry VI board, 270 rings spanning the years 1216–1485;
for the Richard III board, 245 rings spanning the years 1231–1485.

The annual rings on the edges of the boards were all heartwood, but on the back of the Henry V portrait (plate 5) a very small amount of sapwood existed for the years 1495 and 1496, showing that the heartwood/sapwood boundary occurred at about 1495. The narrowness (average width 0·9 mm) of the terminal rings imply that at least twenty-five years were covered by the sapwood. The tree was therefore unlikely to have been felled before 1520 and the likely period for the paintings, that is for the use of the panels, is 1520–25. So even the portrait of Richard III was painted nearly forty years after he was killed at Bosworth.

Plate 5. Henry v. Oak panel, Royal Collection; likely date, by dendrochronology, in the period 1520–25. (Reproduced by Gracious Permission of HM the Queen.)

Tree-Ring Dated Samples of Wood for Calibrating C-14 Ages

At the time of the 8th International Conference on Radiocarbon Dating (held in New Zealand, October 1972) about 800 samples of wood, already dated by their tree-rings, had been analysed for their C-14 content. Because the standard error of the results for each sample is a substantial number of years, the discrepancy between a C-14 age and the corresponding calendar date is not obtained as an exact number of years but only as an approximation. The trends in the 'band', which relates the two quantities graphically, are attributed to fluctuations of the C-14 content of the atmosphere in three cycles of different periods: the first about 10000 years, the second, called secular, about 200 years, and the third about eleven years. (It seems advisable in referring to these cycles to avoid ambiguous terms such as the 'short' or the 'long' term cycle.) European samples of tree-ring dated wood are relevant to all three cycles.

The 10000-year Cycle and the Secular Fluctuations

When the results back to 5400 BC are plotted on one graph, there is a spread in the individual measurements (see the paper by McKerrell) over about eighty years in AD times and over progressively more years with greater ages. This spread implies inaccuracies arising, for example, from (a) a sample being imperfect; (b) the pre-treatment of a sample being imperfect; or (c) the C-14 measurements being imperfect. Other possibilities have been discussed by Clark and Renfrew (1972). When smooth curves are drawn through the various values, some of the 'wiggles', for which evidence has been found by Suess, are eliminated. Yet they may well be real.

Calibrations for the period from 1860 AD to 1100 BC were obtained by Suess (1965) by using 98 dated samples from 24 trees, half of them being from conifers (such as sequoia and Douglas fir) that had grown in the USA, the other half being from sessile oaks supplied by Huber. There was no significant difference between conifers and oaks, both in these results by Suess and also in others between oak from Swiss neolithic sites (Huber and Merz 1963) and contemporary ones on bristlecone pine (Ferguson, Huber and Suess 1966; Mook, Munaut and Waterbolk 1972). At the present time, three sets of buried oaks in Europe are in process of providing calibration curves back to about 6700 BC. The merits of such oaks (*Q. petraea* and *robur*) are the absence of missing rings; their growth at altitudes common to much archaeological material; their large numbers and wide rings; and the advantage that can accrue from the oak being a ring-porous tree (*vide infra*). It is my belief that together they will provide for prehistoric times a more accurate calibration than is at present available from bristlecone pine.

Two of the sets consist of oaks that lie preserved in river gravels in central Europe. At Stuttgart, Becker (1972) has analysed 850 samples, many apparently buried by long phases of floods, while 422 oaks, rescued from extinct courses of the Rhine and the Weser, have been examined at Cologne (Schmidt 1973) together with pollen and other vegetative material from adjacent horizons. The third set are bog oaks recovered in Ireland and under study at Belfast (Smith, Baillie, Hillam, Pilcher and Pearson 1972).

In using tree-ring dated wood to calibrate C-14 ages the assumption is made that the carbon in each annual ring is derived from atmospheric carbon during that very year. There is a slight possibility, through resin migration within a tree, of the presence of carbon synthesised in later years (Jansen 1970), but no contamination by the sap of subsequent years has been detected in oak (Berger 1972). This is in keeping with the finding (Huber 1935) that only the outermost layer of early vessels conducts the sap in oak (*Quercus petraea*).

There is also the possibility of the presence of carbon compounds synthesised in *earlier* years and stored as food in the roots, stems or rays of trees (to aid the continuity of its life). Tubers such as the potato provide an extreme form of this biological process. There have been studies of the release of such food reserves in certain conifers (Krueger and Trappe 1967; Kozslowski and Winget 1964). The process must also be significant with oaks and certain other broadleaved trees, such as ash, walnut and elm, that are termed ring-porous and in which there are large, sap-carrying vessels (seen in plate 1) laid down at the beginning of each year's growth (Huber 1935). They form what is often called the 'early' or 'spring' wood and must derive their food from reserves built up in *earlier* years because it was noted by Priestley, Scott and Malins (1933) that the vessels are formed simultaneously with the opening of the buds over the whole tree.

The effect of this process on the calibration of conventional C-14 ages from results on tree-ring dated wood will depend on its relative magnitude. This will be higher for aged trees and ones of slow growth. In oaks over 200 years old we find that the early wood usually accounts for well over half the width of each annual ring (plate 2)—a much larger proportion than during the vigorous growth associated with the first hundred years of growth. So it would seem advisable when using European oaks to calibrate C-14 dates, to select wide rings from the earlier years of growth. In samples of bristlecone pine, this process may well be significant due to their great age and the narrowness of many of their rings.

The Eleven-Year Cycle

Fluctuations of this duration in the atmospheric $^{14}C/^{12}C$ ratio were anticipated from the existence of the eleven-year cycle of solar activity. However, its detection and the estimation of its magnitude is relatively difficult as it needs wood of adequate weight confined to a single year's growth, or samples of wine, grain or other seeds in which the carbon is synthesised in a single season.

If the magnitude is small, say 0·5 per cent or less, the accuracy of many C-14 datings carried out on samples of twigs, antlers or grain that cover only one season would not be affected significantly: the many C-14 determinations on dated samples that contained a few annual rings would be even less affected. But if the magnitude is as much as 3 per cent (corresponding to $\pm$120 years on C-14 measurements) such material and samples would be subject to an additional uncertainty with a standard deviation of up to $\pm$60 years.

At present, the magnitude and hence the extra uncertainty of a C-14 age derived from a single-season sample is in doubt. Suess (1965) concluded that the variation was less than 0·5 per cent in the twenty-year period 1868 to 1888 from 15-gram samples of tree-ring dated wood taken from a fir that had grown in Oregon. However, at Glasgow, Baxter and Walton (1971) from samples of wines, whisky and flax seeds, suggested a magnitude of about 3 per cent in the first half of this century, and later results (Baxter and Farmer 1973) gave a similar magnitude (2–3 per cent) with 40-gram samples of oak heartwood dated to the years from 1829 to 1865. These came from the butt of a tree felled in 1970 in the Forest of Dean in Gloucestershire, the oak selected having in its early years of growth a series of wide annual rings, mainly between 2 and 5 mm wide (as are those shown in plate 1).

Wood with ten or so years' growth will largely avoid any uncertainty from this affect.

Dating Wood by C-14

As my own archaeological studies (medieval timber-framed buildings in North Berkshire) have led me to submit samples for C-14 measurements, I have experience of the problem of converting what are called raw or conventional C-14 ages into meaningful results with reference to the site under study (Fletcher 1968, 1970). Like others, I have at times unconsciously increased the uncertainty of the result by submitting a single sample or several too low in weight. On the other hand I have avoided submitting samples contaminated with extraneous carbon compounds or those in which there is a considerable uncertainty in the growth allowance.

It is a mistake to believe that C-14 ages for wood used in the last 2–3000 years cannot be easily and relatively accurately converted to meaningful calendar dates. Yet in archaeological reports many are still only given in the form received from the radiochemical laboratory, e.g. as 1150$\pm$50 BP. The situation is different with certain prehistoric material where there is still uncertainty as to what correction to apply for calibration.

If Roman and medieval archaeologists are not to be misled, often by a hundred years or more, it is *essential* with wood (a) that appropriate samples be sent for C-14 analysis so that the uncertainty of the C-14 age is kept low; and (b) that the ages reported from the analysis are corrected. The ages reported (often now designated bp, rather than BP, for 'before present') are usually already corrected for the $\delta^{13}C$ value of the sample. They include the uncertainty from the counting, quoted as one standard error (1σ). The uncertainties given in *Radiocarbon* for wood up to 2000 years old, a period to which the remainder of this section is confined, vary from about ± 40 to ± 120 years.

The Corrections

Corrections need to be made for three items. These were discussed by Berger and Horn (1970) and are listed together with their magnitudes in table 3.

Table 3. Corrections to raw C-14 dates for wood of Roman and Medieval times (1–1500 AD).

Correction for	years	Magnitude: maximum	minimum
1. C-14 half-life	earlier by 0 to 60	older samples	recent samples
2. Calibration for atmospheric variation of $^{14}C/^{12}C$ ratio	earlier by 110, to later by 70	15th century, 9th & 10th centuries	1240-1380 AD
3. Growth allowance of tree	later by 0 to 200	slow grown and old trees	twigs, small branches, young trees, outer annual rings
Items 1 to 3 together	earlier by 125, to later by 270		

C-14 *half-life*. All C-14 ages reported in *Radiocarbon* have to be given in the conventional terms of the now archaic half-life of 5568 years. The generally accepted value is 5730 ± 30 years, that is 3 per cent higher, and so the conventional or raw ages must first be increased by this percentage. The time is bound to come, as with the change from the Julian to the Gregorian calendar, when the results from C-14 laboratories will cease to be presented in terms of the 5568 half-life. Meanwhile the correction is easy to make. For example, an age of 1155 bp (795 ad) becomes 1190 years before 1950, i.e. 760 AD.

Calibration for atmospheric variations of the $^{14}C/^{12}C$ *ratio*. The results published by Suess in 1965 on tree-ring dated samples of US conifers and European oaks were the first to provide a basis for this correction. The results, in the form of a curve relating the C-14 age to the true or 'dendro' age, were extensively applied by Berger and Horn (1970) to samples of wood from medieval buildings in Europe. Suess extended his measurements and revised the curve in 1970 (Suess 1970).

Later calibration tables have included results from ring-dated bristlecone pine. They cover a much longer span of years but for AD times are not necessarily more correct than those of Suess. Two, in general use, that apply to ages (or dates) based on the 5730 year half-life are: (a) by Ralph, Michael and Han (1973), with corrections at 10-year intervals; (b) by Damon, Long and Wallick (1972), with corrections for 25-year intervals. Although the differences between the two for the AD period sometimes extend to 50 years, they are generally within 20 years of one another.

Growth allowance. This allows for the growth (in years) between the annual rings contained in the sample analysed and the ring nearest to the bark, which is either the actual year or the year before the tree was felled (depending on the time of year at which the operation was carried out). With wattles or branches of trees with few annual rings present, the allowance is negligible, while for timber, such as oak, from long-lived trees it is important. Inspection of wet oak timbers recovered from archaeological excavations will often show, by the spacing of the grain, whether the trees were slow or fast grown. When a section is cut, it will usually be even more obvious.

For old trees it is particularly desirable that tree-ring analysis should *precede* the selection of samples for C-14 measurements so that samples with known growth allowance (which could vary from a few years up to more than 200 years) can be selected.

Our work on the oak planks from the North Elmham well (Wade-Martins, Fletcher and Switsur 1973) illustrates a typical procedure. Ring-width measurements were made on 2-inch sections cut from five planks, each with nearly 200 annual rings. The good agreement between their ring-width charts indicated that they came from the same tree. All had some sapwood so the felling year could be estimated, on an arbitrary scale of years, to an accuracy of one or two years, and the growth allowance for any group of adjacent annual rings calculated. Four pieces, each containing about 25 grams of air-dried wood, were cut from the sections and submitted for C-14 analysis. Each section was of heartwood, which is less prone than sapwood to insect contamination, and covered ten to fifteen annual rings. The growth allowances (respectively 171, 116, 82 and 24 years) were chosen to cover a wide range.

For the different samples of timber that have been used during the last two thousand years it is apparent from table 3 that the relative importance of the three corrections will vary from one sample to another. In a few cases one correction may almost cancel the other two, but such occurrences are no justification for ignoring the relevance of these corrections or, with oak, for assuming that pieces that appear to be near the sapwood need no correction for growth allowance.

Sampling and the Uncertainty Term

For historic times, a high uncertainty, such as ±100 years in the C-14 age, or ±130 years in the calendar date derived from it, makes the result of limited value. Though sometimes helpful as a preliminary indication, within 2–300 years, of the date of an artifact or site, it means that more accurate dating, either by C-14 analysis of several well-selected samples or by dendrochronology or some other evidence, will be necessary. Table 4 gives the three factors which contribute to the uncertainty of a C-14 age.

Table 4. Sampling and the uncertainty of C-14 ages of wood.

1. Interference from background radioactivity and cosmic rays with the counting of C-14	The uncertainty from this can be reduced (a) by providing samples of adequate weight, (b) by counting for a longer time, (c) by providing more than one sample.
2. Contamination by extraneous carbon compounds *either* much older, e.g. from coal, oil; *or* much younger, e.g. humic acid	(a) Difficult to detect. (b) Chemical pretreatments aim to remove contaminants but may fail to do so completely. (c) Removal of contaminants from buried timber made easier if samples are not allowed to dry out. (d) Sometimes indicated by duplication of samples.
3. Errors in manipulation during the C-14 analysis	Unpredictable

Interference with counting of C-14. The uncertainty quoted by a C-14 laboratory will certainly include this factor and, for clarity, should be *limited* to it. In the past this has not always been so. For a given time of counting, the uncertainty will be larger when the counts from the C-14 are low. This can occur either because the amount of carbon is smaller than required to fill the counting chamber, or the carbon is old in comparison with the half-life, 5730 years, of C-14. The second alternative is scarcely significant for historic material.

To avoid the waste of time and money involved in counting for longer times, the archaeologist submitting a sample for analysis should ensure there is enough to provide, after pre-treatment, carbon of adequate weight. While waterlogged samples of medieval timber contain only 15–10 per cent of carbon, dry timber from buildings contains about 45 per cent.

Uncertainties of the order of ±30 years from this factor have been achieved with three or four samples, the $\pm$ term being reduced by the square root of the number of samples.

Contamination. It is not easy to separate all contaminants so that their residue amounts to, say, less than 0·1 per cent of the carbon from the wood. Some pre-treatments, e.g. isolation of the cellulose, often drastically reduce the weight of carbon, in itself a disadvantage.

In samples of wood from wet sites there may be both recent carbon from humic acid and animal products retained by ion-exchanges processes, as well as ancient carbon derived from petroleum and coal products. Similarly, the dry wood of building timber may have been permeated in the past by phenols and cresols liberated as vapours by the combustion of wood, or recently by preservatives sprayed on them. There is also the possibility of recent carbon from insect attack (plate 3).

In some pre-treatments volatile organic solvents are used to remove some of the organic contaminants, but Jansen (1972) found that the complete removal of the solvents can itself present a problem.

Extraneous carbon compounds, as well as the third factor mentioned in table 4, errors in manipulations, are likely to have produced the conflicting results that are known to have occurred (Jansen 1970). When one out of several samples from the same source of material gives an apparent C-14 age widely (in terms of the counting standard error) different from the others, it is usually better to discard it as being due to contamination or a manipulative error, than to include it and enlarge the uncertainty term of the mean.

Examples and Conclusions

In the work already mentioned on the oak planks from the Saxon well at North Elmham (Wade-Martins, Fletcher and Switsur 1973), four samples with widely different growth allowances were submitted for C-14 analysis. The C-14 ages were corrected, as shown in table 5. The uncertainty of the mean, 832 AD, was ± 30 years. This is made up of ± 20 years from the counting error, i.e. ± 40 years divided by $\sqrt{4}$; and ± 22 years from the uncertainty in the calibration correction.

Similarly, from the C-14 analyses by Berger at Los Angeles (Berger and Horn 1970), some corrected dates with an uncertainty of only ± 25 to 40 years have been obtained for parts of timber-framed buildings from samples collected either by Professor Horn and Frau Giertz in England and Normandy, or by myself in North Berkshire. In these examples—the Middle Littleton Barn (Berger and Horn 1970) and the Tractor Shed at Long Wittenham (Currie and Fletcher 1972)—three or more contemporary samples were analysed to achieve results of relatively high accuracy and low uncertainty.

For wood, it would seem more appropriate to use the equipment in C-14 laboratories for investigations of this type than on a greater number of *single* samples, particularly if the results for such single samples are published without any correction.

Table 5. Corrected C-14 ages for 4 samples of oak planks from North Elmham.

	Q 1043	1041	1042	1040
C-14 age	1315	1266	1220	1172 bp
Correction for C-14 half-life	+39	+38	+37	+35 years
Calibration for secular variation	−61	−62	−65	−62 years
Correction for growth allowance	−171	−116	−82	−24 years
Corrected age	1122	1126	1110	1121 BP
Calendar date	828	824	840	829 AD

It would seem advisable for an archaeologist aiming to use wood to date a historical site:

(a) To determine the limits of the period (half-century or century) that would provide him with new and meaningful information.

(b) To enquire from a radiocarbon laboratory or from a suitable consultant
   (i) if a result within the limits desired is feasible;
   (ii) if so, if dendrochronology should precede the selection of samples;
   (iii) the number of samples that need to be analysed, the weight required for each, and other necessary technical details about them.

(c) Either to forego C-14 analysis on the grounds of it being uninformative for the material concerned (or too costly), *or* to provide samples of appropriate number, weight and quality.

(d) To correct the conventional C-14 ages reported.

Irregularities in the Dendrochronological Calibration Curve

J. H. OTTAWAY & BARBARA OTTAWAY

This paper has nothing to say about the magnitude of any corrections that may have to be made to conventional C-14 dates, although it does suppose that a calibration curve exists. It is solely concerned with the reality or otherwise of short-term deviations from a linear or smooth curve.

The calibration curve drawn by Suess (1970) has a large number of wiggles, which caused dismay among archaeologists because in some regions of the curve it was possible to obtain several corrected dates for one conventional date. We have shown (Ottaway and Ottaway 1972) that strict application of this curve to assemblies of archaeological dates led to distortions of the distribution of dates within some assemblies that appeared to us quite unacceptable, caused by the very rapid changes in slope of the curve around the wiggles. For this reason we thought that most of the wiggles could have no real existence. Several contributors to the Nobel Symposium on Radiocarbon Dating (e.g. Damon, Long and Wallick 1970) drew attention to sound geophysical reasons for expecting some short-term fluctuations in the rate of atmospheric C-14 production, and in the light of this evidence we were rather cautious about concluding that there are no wiggles at all in the calibration curve. However, several recently-published curves, for example those of Wendland and Donley (1971) and Switsur (1973), are almost or completely smooth. To establish the form of the curve with confidence it seems clear that a method of locating short-term variations is needed which is independent of dendrochronological measurements. Suess pointed out to us that one could use for this purpose our method of 1972 in reverse, as it were. Thus if a set of conventional C-14 dates, which should *a priori* be evenly spaced, is found to be distorted in its distribution in time, this is evidence for a wiggle in the curve relating C-14 time to absolute time, and its centre would locate the position of the wiggle. A diagram will make the argument clear (figure 1).

If a series of objects is evenly distributed in absolute time (ordinate), and if there is a hiccup in the curve relating this time to conventional C-14 years (abscissa), the frequency distribution of the *apparent* dates of the objects will be irregular. Some objective test of the distribution should then be able to pick up the irregularity.

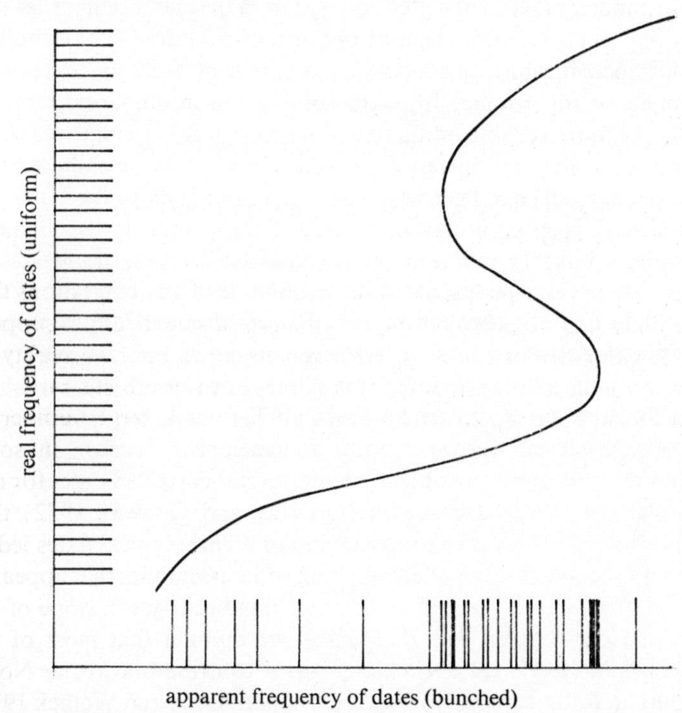

Figure 1. Effect of a 'wiggle' in the curve relating absolute and C-14 dates on the apparent distribution of C-14 dates (abscissa). The objects giving the dates are assumed to be uniformly distributed in real time (ordinate).

In practical terms the only way of obtaining an even distribution of dates in absolute time is to use a very large number, in the expectation that local variations in intensity stemming from the interests of a single investigator will be swamped. To this end all the C-14 dates which have appeared in the journal *Radiocarbon* from volumes 1 to 15 (part 2) were punched on to computer cards. As this was a fairly considerable undertaking, rather few judgements were made about the entries. 'Archaeologic' dates were collected separately from 'geologic' dates, but the classification of the submitter was used (for example, the first Somerset Levels dates were submitted as geological, while the later ones were described as archaeological). Deep-sea samples were excluded, but no attempt was made to exclude shell or tufa samples, and samples from both hemispheres were taken together, although it is possible that the atmospheric C-14 distribution has not always been uniform all over the globe. These omissions may have caused some wiggles to be obscured, but it was thought more important to include the maximum number of dates in a preliminary test of the method, in order that the most important criterion, that of uniform distribution, should hold.

In all instances in which two separate measurements were made from the same sample, as for instance from the soluble and insoluble fractions of peat, only the more reliable estimate was taken, but no attempt was made to remove replicates, i.e. instances in which two or more samples (for example, of charcoal) had been taken from the same layer of the same site. This could have been an important source of error since, in the archaeological series, about 15 per cent of the dates occurred as duplicates or replicates. However, small-scale tests showed that removal of all the replicate dates had no effect on the distribution of statistically significant wiggles. A more serious source of error was found to be the tendency of scientists to round off the estimates of the dates in an inconsistent manner.

About 26000 dates up to 50000 years BP were collected, some 13800

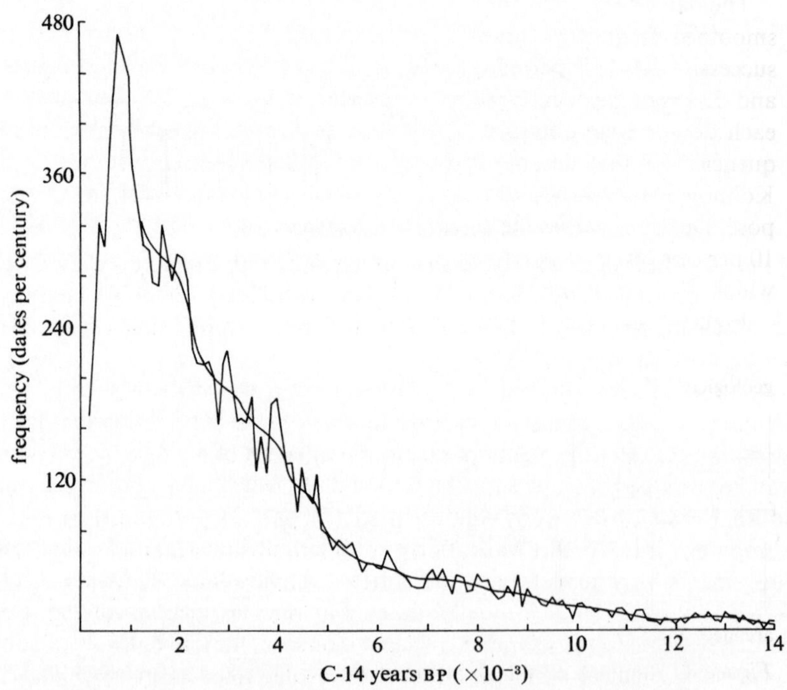

Figure 2. Smoothed frequency curve for all dates in vols 1–15 of *Radiocarbon*.

archaeological and 12500 geological. Unfortunately most of the archaeological dates are concentrated in the first three millennia BP (figure 2), largely because of the intensive work that has been done on American Indian pre-history, so most of our studies were made on the geological data, which are fairly frequent up to 14000 BP (figure 3), or on the combined data up to 40000 BP (Ottaway and Ottaway 1974).

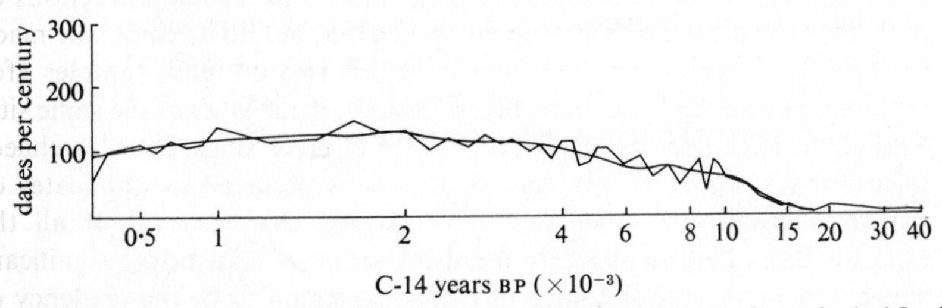

Figure 3. Smoothed frequency curve for geological dates in vols 1–15 of *Radiocarbon*.

The dates were first sorted into ascending order of magnitude, and then smoothed frequency curves were computed from the frequencies over successive 200-year periods. These smoothed curves are shown in figures 2 and 3. From them interpolated estimates of the expected frequency for each decade were obtained. These were compared with the observed frequencies for that decade, in short overlapping time intervals, using the Kolmogorov–Smirnov test, and a computer program written for the purpose. Deviations from the expected frequencies were collected at the 1, 5 or 10 per cent levels of significance, but most attention was paid to deviations which were registered in several of the overlapping test sequences.

Figure 4 shows the regions in which wiggles were found to occur. The

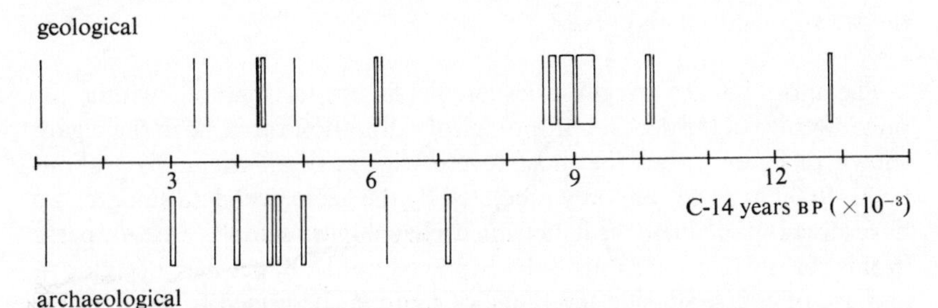

Figure 4. Regions in which statistically significant irregularities in C-14 dates occur in the period 0–1400 BP. The width of the boxes is proportional to both duration and intensity of disturbance.

upper part of the diagram gives the results from the geological data alone, the lower part from the archaeological data alone. Although there are several regions where wiggles are only predicted by one of the sets of data, the general agreement is good. This suggests quite strongly that the method functioned satisfactorily, because several of the laboratories analysing geological samples (notably the Geological Survey of Canada) do not measure archaeological samples, and to some extent *vice versa*.

Figure 5 shows the major wiggles in the period up to about 7000 BP superimposed on the recently-published MASCA curve (Ralph, Michael and Han 1973), and figure 6 shows the period from 3500–5000 BP in more detail. The agreement here is very good indeed.

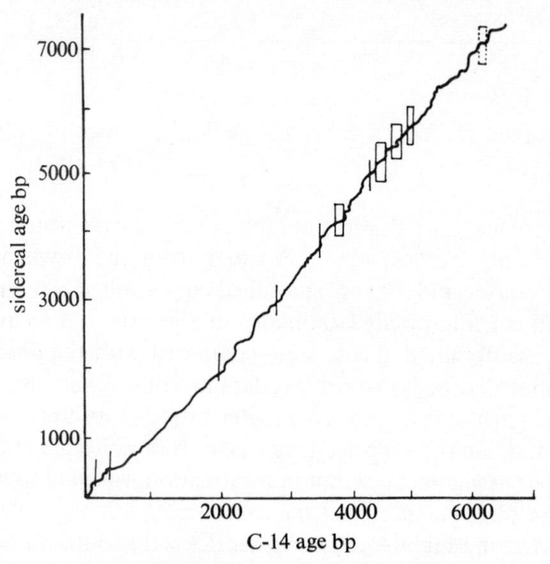

Figure 5. Major irregularities in the period 0–7000 BP, superimposed on the MASCA calibration curve.

The most intense irregularities predicted by our method, within the present range of the dendrochronological calibration curve, lie in the region shown in figure 6, but the most severe wiggles of all occur from about 8500–9000 BP. They are only predicted by the geological data since, as we have already explained, the data from archaeological samples are too sparse in this region. They are thus only displayed on the upper part of figure 4, and are of course outside the range of figure 5. It seemed to us plausible that this intense disturbance must coincide with an equally intense climatological event. In order to see whether there is any factual support for this hypothesis, we compared the C-14 dates of all the wiggles, large and small, that we could locate, with the C-14 dates of the maxima of glacial recessions over the last 20000 years BP, recently published by Bray (1972). There is very good agreement between the two sets of dates, as table 1 shows; thus without in any way speculating on cause or effect, it appears probable that irregularities in the relation between C-14 age and tree-ring age reflect short-term changes in the earth's atmosphere and in the oceans, and particularly in their temperature.

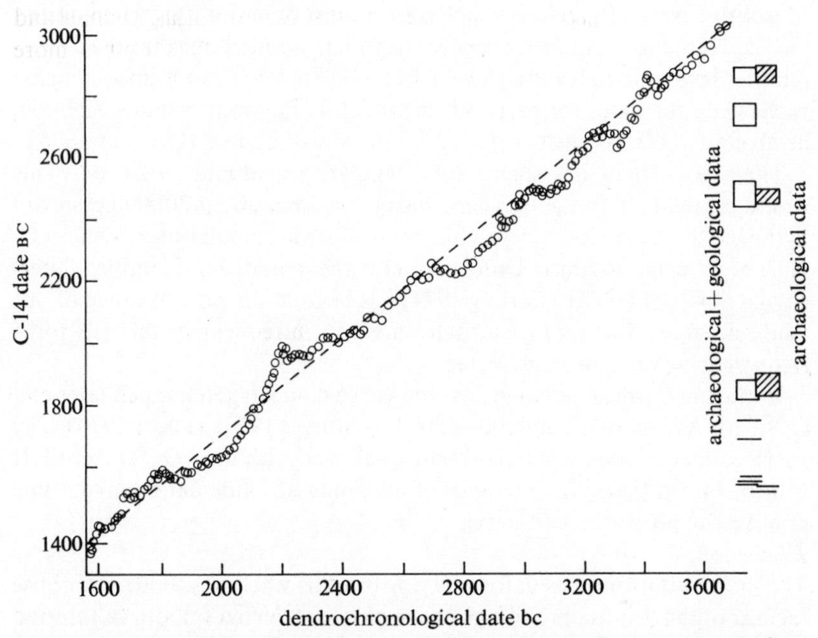

Figure 6. Correlation of irregularities predicted by the frequency technique with wiggles in the MASCA calibration curve, over the range 1800–3600 bc. Note the good agreement at 1800, 2200 and 3400 bc.

The biggest climatological disturbance known to us in the recent past is the ending of the last Ice Age. If this is associated with the intense wiggles just before 9000 BP (figure 4) we have in principle a way of assigning an absolute date to this relatively short set of C-14 dates, and so in effect of extending the C-14 calibration curve considerably beyond its present range.

The end of the Ice Age was of course not a single event, but the most rapid temperature fluctuations appear to be associated with the relatively short-lived Allerød period (table 2). Unfortunately, there is at present no generally accepted absolute date for this or any other part of the transition period. Figure 7 shows three different estimates. The figure that seems to us most reasonable comes from the $^{18}O/^{16}O$ ratio profiles in the Greenland and Antarctic ice cores (e.g. Johnsen, Dansgaard, Clausen and Langway 1972). These show the Allerød change to be completed by about 11 000 bp. It is of interest that if the dendrochronological curve were extrapolated backwards without further change of slope, the period 8600–9000 BP would correspond to 10 500–11 000 bp, registering almost exactly with the ice core date. If this assignment can be made, it is satisfactory from the archaeological point of view, because it would allow a greater time range for the European Mesolithic period in regions which were almost or entirely glaciated during the Ice Age. J. B. Campbell (pers. comm.) has pointed out that the uncorrected Mesolithic dates are bunched closely together in an unconvincing way, while the dates for early Vinca, which is far from being Mesolithic, lie around 7000 BP when corrected by the MASCA curve (Ottaway 1975).

There is a small but significant irregularity at about 29 000 BP. This would correlate with the transient warm period at about 30 000 bp shown in the ice sheet profiles, for which there is other accumulating evidence (D. Harkness, pers. comm.). Unfortunately, the tentatively identified interstadial at 42–43 000 BP (Girling 1974) lies beyond the present range of our method, since even geological dates are too infrequent in this era for a frequency curve to be constructed.

It must in fairness be noted that the varve data suggest a much later end to the Ice Age, perhaps at 8 000–8 500 bp (Stuiver 1970; Tauber 1970). The varve counts do not seem to correlate at all well with the $^{18}O/^{16}O$ record. It is to be hoped that other methods of assigning absolute dates may in time remove the present uncertainty.

Discussion

The most important reason for carrying out this work was to identify those regions of the dendrochronological age plot for which a smooth calibration curve might be constructed. It was hoped that these regions would be large ones, and that the intervals in which it would be unwise at present to construct a simple curve would turn out to be limited; fortunately this seems to be the case.

The uncertainty that attaches to a single estimate of a date of an object by the C-14 technique is part of the estimate, and must not be lost sight of. This uncertainty can never be smaller after the correction than before, but with a good calibration curve the increase may be quite small. It must be pointed out that the essence of a calibration curve is to convert values of a dependent (or measured) variable into values of an independent (or absolute) variable. In order to estimate the confidence limits of the independent variable, it is necessary to have, in addition to the confidence limits of the original measurement (e.g. its standard deviation), the general variance of the dependent variable in the calibration graph. To those who, like

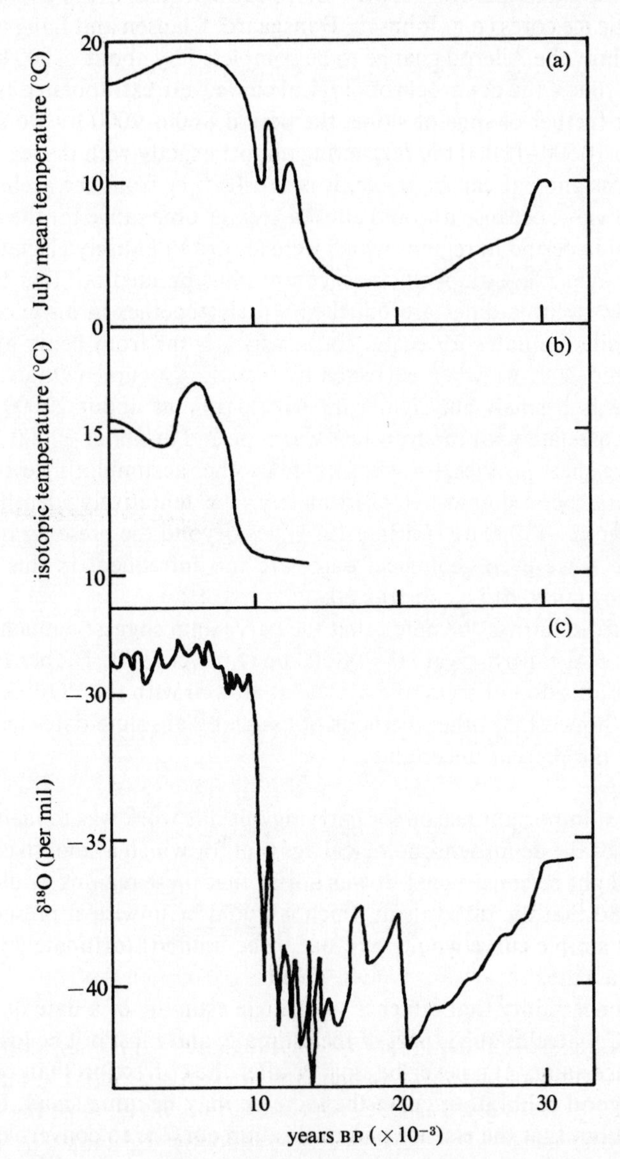

Figure 7. Estimates of the end of the last glacial period: (a) from palaeobotanical estimates of the July mean temperature in the Netherlands; (b) from estimates of the surface temperature in the Mediterranean; (c) from the isotopic oxygen ratio in the Antarctic ice sheet. Notice that the steep fall in (b) begins 2–300 years before that in (a) or (c).

Table 1. Significant irregularities (C-14 years) in the frequencies of the geological C-14 dates.

Date bp	Significance level	Date of temperature maximum
890	*P*<0·1	900
1620	0·05	1600
3180	0·1	—
3350-3390	0·1	3300
*		3800
4200-4210	0·01	—
4460-4470	0·05	4400
—		4500
4660-4690	0·05	—
—		5800
—		5900
6000, 6060		—
6550-6640	0·05	6500
†		7100
8600-8800	0·01	8500
8910-9050	0·01	8800
—		9600
—		9700
10140-10190	0·05	—
10990	0·05	11110 —
11980-12020	0·05	11800
		12250
12590-12640	0·1	—
—		13400
—		14100
14640-14790	0·1	14700-15500
		16100, 16800
—		18000, 18300
		19100, 20300
27430-27490	0·1	—
27630-27660	0·1	—

Estimates of significance are conservative.
* Irregularity at 3770-3810 yr bp in combined data (*P*<0·01).
† Irregularity at 6910-7020 bp in archaeological data (*P*<0·05).

ourselves, believe that dendrochronological ages are, if not absolute, at least more closely related to absolute age than are conventional C-14 dates, the latter are the *dependent* variable in a calibration curve, while the tree-ring ages are the *independent* variable. Consequently the tables of standard errors of tree-ring ages published by McKerrell in this Symposium, although interesting, are of no help in estimating confidence limits, because they provide information about the variance of the wrong variable. There are good statistical reasons why a calibration curve is better than a series of fifty-year averages for converting from a dependent to an independent variable, and although the estimate of the true statistical uncertainty of the converted estimate is quite complex (Colquhoun 1971), Switsur (1973) has quoted a simplified version for use with his own calibration curve which is perfectly acceptable. It appears desirable to us that a similar curve, which takes into account the regions predicted by us where caution ought at present to prevail, should be brought into use as soon as possible. It has been shown that for groups of dates defining cultures estimates of uncertainty for each individual date are not necessary (Ottaway 1973, 1974), but the contribution of A. Snodgrass to this Symposium reminds us that crucial olive trees cannot necessarily be replicated on demand.

Table 2. Periods of temperature fluctuation at the end of the last Ice Age.

Pollen zone	Climate	Conventional C-14 years BP
Pre-Boreal	warm	
		10150
Younger Dryas	cold	
		10950
Allerød	warm cool warm	
		11750
Older Dryas	cold	
		11950
Bølling	warm	
		12350
Oldest Dryas	cold	

To lay stress on the statistical arguments involved in C-14 dating is not to argue that archaeology should be subservient to statistics, but those who use a powerful tool in the furtherance of their research should not fail to use it properly (compare, for example, the frequent misuse of the microwear technique for flint tools; Keeley 1974). The uncertainty of a single estimated date is very real to those who have to spend much time on the accurate determination of low levels of radioactivity.

An unexpected result of this research is the very good alignment between the self-predicted wiggles in the C-14 frequency curve, and the glacial temperature record. We realise of course that our analysis has been relatively crude. We might have taken more trouble to separate pelagic from terrestrial samples, to deal with the Northern and Southern Hemispheres separately (since the ice-sheet profiles are not identical for Arctic and Antarctic), and we might, with considerable effort, have removed all the replicate samples which may have distorted the analysis. There was not, however, sufficient time available to us for these refinements, and we are content to have shown that the method works in principle, and that probably all wiggles of any consequence have been identified. It is encouraging that a major cause of distortion, which we did have to circumvent, was not any of the factors just referred to, but the habit of physical scientists of rounding off numbers in a non-random way when they are near a terminal -50 or -00.

It is of course possible that Bray (1972), whose figures we used for this comparison, overstated the agreement between the different estimates whose results he collated for his table of glacial retreats. It is certainly desirable that other evidence of short-term climatic variation (e.g. marine transgressions, I. Morrison, pers. comm.) should be tested against our predictions. There seems nevertheless considerable hope that C-14 wiggles can play a part in unravelling the complex history of climatic variation.

An Outsider's View of C-14 Calibration

A. M. SNODGRASS

Some far-reaching claims have been made in the past few years for the effects of the calibration of C-14 dates on the prehistory of Europe. 'All our existing text-books' we have been told 'have to be scrapped, and completely rewritten in these terms' (Renfrew 1973b, attrib.). Let us pause for a moment before we consign our copies of *Ancient Europe* to the shredding machines, and consider why we are being asked to do this. What has taken place is a chronological adjustment, or rather a proposed chronological adjustment. Now chronology is a small hinge on which great doors may swing: a slight adjustment at a vital point, and the whole development of the interpretation of major cultural processes may be suddenly and violently reversed. Because of this, we must be especially careful that the adjustment is really necessary: and we need accurate and reliable tools to execute it. Most authorities seem to agree that C-14 dating is, in the present state of knowledge, a slightly blunt instrument (more so than is suggested by the conventional notation used for the results of laboratory tests), while calibration, for all its illusory effect of refining the instrument, in fact blunts it further by widening the margin of uncertainty in many cases. On an excavation, the circumstances that can take an isolated archaeological find thousands of years away from its true chronological context are not altogether different from those producing an isolated C-14 date that misrepresents its chronological context by several hundred years. From all this it must surely follow (though I must emphasise that, as my title suggests, I speak as no expert in these techniques) that we need a solid body of C-14 evidence on each side, and a well-proven scale of calibration, before we can safely invert the supposed chronological relationship of any two cultures.

Now one of the world's leading centres of research into both C-14 dating and calibration is at the University of Pennsylvania, and it was therefore interesting to read the comment of a professor of that university (Muhly 1973, 411–12) that 'scholarly caution is the proper attitude at the present time', to recent proposals for revising the chronology of the Wessex and other northern European cultures. I can only agree, and indeed go further in calling for the exercise of a more positive quality: what one might call scholarly rigour. (Again, I hasten to add that I do not see myself doing anything more than aspiring to that quality.)

Any discussion of the later prehistory of Europe, even in these days of independent development, is likely to take some account of the Aegean. In particular, it is hard to imagine a view of the Bronze Age of Europe which did not admit of *any* connection with the Aegean Late Bronze Age. I hope it will be enough to mention such names as Nienhagen, Perushtitsa, Ørskovhede, Čaka and Pelynt, each of them places where objects believed to be either locally made under strong Aegean influence, or actual Aegean imports, have been found in archaeological contexts. (We have perhaps now been frightened away from adding names like Rillaton and Bush Barrow to the list.) The currents of influence, of course, run from north to south as well as from south to north, and I think it will prove equally impossible to destroy completely the validity of either. It might have been expected, therefore, that any reconsideration of the absolute chronology of Bronze Age Europe, on the basis of calibrated C-14 dates, would take account of such dates from that area of Europe from which almost all the evidence for the traditional chronology had been drawn, the Aegean of the Late Bronze Age. But as far as I can see this has not been done, except very briefly by McKerrell (1972, 295). Instead, the favoured method seems to be to base conclusions on a comparison of calibrated C-14 dates from one part of Europe with historically derived ones from another. This seems a fundamentally unsound method, and a quite unnecessary one since a body of material is available on both sides for a truer comparison.

It may be, perhaps, that the oldest barrier in archaeological studies has once more loomed up here: the mutual distrust between the practitioners of true prehistory and those of the archaeology of literate civilisations. It may be that the C-14 evidence from the later Aegean is tacitly dismissed as being the casual by-product of the work of people with their minds on other things. But I hope to show that the dates from the later Aegean series show, both in the sampling and in the laboratory stages of processing, internal evidence of competence that does not significantly differ from the level of competence shown elsewhere. After all, the samples have mostly been processed by the same people, some of whom have also helped to pioneer the work of calibration. Even if this were not so, I think it would be quite unjustified to assume that dates from central, northern or western Europe were generally immune from the weaknesses attributed to the Aegean ones. Table 1 is a select list of twenty-one Aegean Late Bronze Age C-14 dates, chosen for their apparent association with one or other of the major and well-established horizons of destruction in later Aegean prehistory to provide a closer check with historical dates. I have omitted a handful of other available dates from Knossos and Mycenae because they seem to me less closely related, if at all, to these prominent landmarks.

Table 1. C-14 dates believed to be associated with historically-dated destructions on Aegean Late Bronze Age sites.

Site	Description	No.	Date (5568 half-life)
Thera, ancient building in quarry near Phira	charcoal	L-362	1410±100
Thera, Phira quarries	carbonised tree	P-1401	1456±43
Palaikastro, LM IB destruction	contents of jar	St-1263	1585±70
Palaikastro, LM IB destruction	wooden threshold	St-1264	1560±120
Ayia Irini, LM IB/LH IIA destruction	charcoal	P-1282	1176±48
Ayia Irini, LM IB destruction	charcoal	P-1284	1300±66
Knossos, LM IB destruction	charred wood	P-1356	1014±74
Knossos, LM IIIA destruction with tablets	charred wood	P-1359	1276±43
(Pylos, 'earlier than the palace'	charcoal	P-254	1499±54)
(Pylos, 'earlier than the palace'	charcoal	P-270	1385±41)
(Pylos, 'possibly contemporary with the main building of the palace'	charcoal	P-340	1371±48)
Pylos, Palace of Nestor	squared door-jamb	P-326	1500±58
Pylos, Palace of Nestor	squared beam	P-330	1405±40
Pylos, Palace of Nestor	squared beam	P-329	1306±55
Pylos, Palace of Nestor	squared beam	P-328	1265±57
Pylos, Palace of Nestor	door-jamb	P-332	1185±53
Pylos, Palace of Nestor, destruction debris	charcoal	P-337	1076±53
Pylos, Palace of Nestor, destruction debris	charcoal	P-341	1055±48
Mycenae, Citadel House, destruction level	charcoal	P-1455	1024±49
Mycenae, Citadel House, destruction level	charcoal	P-1456	1085±65
Mycenae, Citadel House, apparently destruction level	burnt beam	P-1457	998±49

Let us begin with the major series of dates from the Palace of Nestor at Pylos. By traditional dating methods it had been concluded that this building had been destroyed at about 1200 BC, after a life that probably did not exceed a century. When, therefore, a series of C-14 dates was published thirteen years ago (Kohler and Ralph 1961, 366–7), ranging at the extremes from 1500±58 (P-326) to 1055±48 (P-341), some initial misgiving might have been felt at the results. But then one looks more closely, and finds that the majority of the dates were either from contexts earlier than the palace, or from the massive, often squared timbers of the once imposing structure. At the end of the series, however, stand two dates of which this was not so. They were taken from small pieces of charcoal from the destruction-floors, and they yielded dates of 1055±48 (P-341) and 1073±53 (P-337): quite close together, and separated by a margin of over a century from the main group of dates taken from large timbers. So from this point of view the series of ten dates from Pylos forms a credible group. At once it was clear that the last two dates were too low and adjustment was necessary: in 1961 by a preference for the longer half-life, today doubtless by calibration. But which calibration? Readings from the published versions of the Suess curve (see most recently Burleigh, Switsur and Renfrew 1973, 316) appear to yield a mean date rather before 1325 BC for these two samples, whether or not one adopts the smoothed curve advocated by Switsur. The MASCA curve, when read according to the authors' instructions, yields a corrected spread of 1300–1110 BC for both dates, and appears to meet the historical case rather better. At this point, it may be observed that the three dates from the destruction-level of the Citadel House at Mycenae (Lawn 1970, 584), which must belong to the same period and could quite possibly be due to the same actual events, give a convincingly close correspondence with the pair from Pylos. If taken as a group of five, they have a reassuring degree of overlap.

But there remains the other issue that I wish to emphasise, namely comparability. The dates are there, and have been for thirteen years in the case of Pylos, for comparison with C-14 dates from further north. If (again in the case of Pylos) a building thought to belong entirely to the sixth of the seven main phases of Mycenaean culture (LH IA-B, LH IIA-B, LH IIIA, B and C) can yield mean C-14 dates ranging from 1500 to 1055 bc, then should not these be taken in conjunction with those from other areas? To mention two such areas, some of the dates from the British and Breton Middle Bronze Ages cited by Renfrew (1968, 278–9), and the one from the bottom of a Y-hole in Stonehenge III B-C (1240±105; *ibid.* 281), lie squarely within the Pylos spread. It will at once be pointed out that the British dates in this group come from phases *after* the periods in which contact with the Aegean had been held to exist. But the rejoinder is equally obvious: the Pylos evidence too lies very late within the Mycenaean age, between two and three hundred years after the Shaft Graves of Mycenae. But the issues, especially that of the comparison of like evidence with like, are better exemplified by other Aegean dates.

First, let us in passing note the date of 1276±43 from Knossos (P-1359), from a horizon *probably* to be equated with that of the main destruction of the Palace of Minos in Late Minoan IIIA (Stuckenrath and Lawn 1969, 159). This date is satisfactory inasmuch as it lies substantially earlier than the last group from the Pylos/Mycenae destruction, yet later than most of those from the Late Minoan IA and IB horizons to which we shall come shortly. Calibration would hardly be justified with such an isolated date, but it is perhaps worth saying that many Aegean scholars (although this is a notoriously controversial case) would agree on a historical date in the region of 1375 BC for this disaster.

Next, nearly everyone must by now be familiar with the great Late Bronze Age volcanic eruption of the island of Thera (on which see Luce 1969). The director of the recent excavations there has repeatedly stated his view that the destruction of the settlements on the island, which may or may not be exactly contemporary with the great eruption, took place at a date close to 1500 BC. He has held this position (e.g. Marinatos 1970, 7–8) in the face of some pressures to lower the date by fifty years or so, pressures that arise mainly from the desire to connect the disaster with the destruction of many sites on Crete, at a date which should be close to 1450. No Aegean scholar would entertain a date much *earlier* than, say, 1520 BC. Now there are some C-14 dates available for the Thera destruction, although only two are known to me at the time of writing: first, a piece of charcoal from a building buried under the pumice of the Thera quarries was sent for testing by Professor A. G. Galanopoulos as long ago as 1956 and produced (admittedly only after a second test) a date of 1410 ± 100 (L-362; Galanopoulos 1958); secondly, and perhaps more significantly, a young olive-tree, actually growing when carbonised by the final destruction of the settlements, gave a date of 1456 ± 43 (P-1401; Marinatos 1968, 55–6). These dates, again, hang together well, although one would really like more of them.

Before we consider the possibility of calibrating them, there is an important implication that they have in traditional archaeological terms. The destruction of Thera is tied to the great series of burials in Shaft Graves at Mycenae, as has been shown by Marinatos in an unusually precise way. The product of a single workshop, probably the hand of an individual vase-painter, the 'Master of the Swallows', has been recognised in a group of works, including one from a late Shaft Grave at Mycenae and one from the destruction at Akrotiri on Thera (Marinatos 1969, 68). Marinatos indeed suggests that the artist may have perished in the volcanic destruction. Be that as it may, the evidence shows a close chronological equation between the Thera disaster and a late phase of the Shaft Graves at Mycenae; and this equation is now reinforced by the identification of several pieces of Theran pumice on a Greek mainland site, Nichoria, in excavated levels of the immediately post-Shaft Grave phase (Rapp, Cooke and Henrickson 1973). I bring the Mycenae Shaft Graves into the discussion because they form a significant epoch in Aegean prehistory from the point of view of northern links, if only because of their distinctive amber finds. It is not easy to suggest a precise absolute duration for this great series of burials, but when the combined evidence of Grave Circles B and A is taken into account, few would argue that it covers less than 150 or 200 years.

With this in mind, let us approach the calibration of the Thera dates, however tentatively. It has been argued by some that the actual volcanic eruption could have come up to fifty years after the desertion of the settlements at the end of the Late Minoan IA phase near 1500 BC, but naturally it cannot have come *before*. The corrected spread on the MASCA curve for the first date (L-362) is c. 1750–1510, which is credible enough; but it was the second specimen (P-1401), taken from an olive-tree growing till the very moment of destruction, which seemed to offer the more accurate indication of date, and here the corrected spread of c. 1730–1600 looks a little high. I am not going to draw any far-reaching conclusion from such isolated evidence; but the result of a cross-reading on the original Suess curve, giving a mean apparently in the eighteenth century BC for the second date and allowing almost no chance of correspondence with the historical date, was enough to raise certain doubts about this calibration for at least one part of Europe. The unsettling implications of this evidence have indeed already been commented on in print (Page 1970, 36).

But again, the case is relevant to the other issue which I have been airing: that of the necessity to compare like evidence with like. If we take it as a working hypothesis that we have what roughly amounts to a terminal date for the Mycenae Shaft Graves in the region of 1450–1400 bc in C-14 years, then it is this date, and not the estimated historical ones, which must be used for comparison with other cultures dated by a C-14 chronology; cultures such as Wessex, for example. One of the latest and most sceptical treatments of the Wessex-Mycenae links hesitates to exclude 'the probability of establishing long-distance connexions' (Briggs 1973, 320). Today of course we have at least four late Wessex dates—Hove at 1239±46 bc, Earl's Barton at 1264±64 and 1219±51, Edmondsham at 1119±45—which on any account must lie far later than our estimated terminal C-14 date for the Shaft Graves. Surely, the fairest inference from the point of view of cross-connections is not that we derive from the MASCA curve a maximum corrected spread for these three Wessex burials of c. 1505–1215 BC (cf. Renfrew 1973c, 223: 'such graves were being constructed until after 1500 BC in calendar years'). No, it is this: irrespective of calibration, and on a straight comparison of like with like, it would seem that the Wessex culture disappears from view, still going strong, at least 150 and possibly 300 years after the *end* of the Mycenae Shaft Graves, from which recently fashionable theories would have dissociated it on the grounds that it too clearly terminated before their *beginning*. I hope that this will discourage premature dogmatism in the future.

One could go on further in this vein. Let me give one more example from the Aegean. I mentioned before the great Cretan destructions of about 1450 BC: in archaeological terms, at the end of Late Minoan IB. From one of the sites then destroyed, Palaikastro, we have two dates from charcoal samples (Engstrand 1965); they are 1560±120 (St-1264) and 1585±70 (St-1263), the former from a wooden threshold and the latter from the contents of a jar, both strikingly high dates even without calibration. By contrast, destruction-levels elsewhere, characterised by pottery of similar date and probably to be associated with the same events, have given dates of a very different order: at Ayia Irini on Keos, 1176±48 bc (P-1282) and 1300±66 (P-1284); and again at Knossos, 1014±74 (P-1356) (Stuckenrath and Lawn 1969, 156–8). The main consequence may well be that our faith in isolated C-14 dates is further shaken. But it is also tempting to point out that the two Palaikastro dates, if looked at in isolation, would bring the Cretan destructions to within touching distance of the period around the construction of Stonehenge IIIA (Renfrew 1968; 1620±110 for an incomplete R-hole (I-2384) and 1720±150 for the erection-ramp of a trilithon (BM-46)), if we were to compare C-14 dates only. Indeed the Palaikastro dates have at least the virtue of lying close together whereas the two Stonehenge dates, as has been noted, fall in the reverse order of that expected (the incomplete R-hole of Stonehenge II should precede the trilithon of IIIA), and the Cretan destructions mark approximately the end of the *third* Mycenaean phase in mainland Greek terms, Late Helladic IIA. In such a situation, it is better not to compound the uncertainty generated by these dates by attempting to calibrate them; the implications should be clear enough already.

The debate about the choice of calibrations will continue, and I do not think that these Aegean dates have a sufficiently sound scientific basis to make a decisive contribution to the question. Indeed better authorities than I (such as MacKie *et al.* 1971) have sketched in the areas of doubt—over the constancy of solar radiation, over differences of longitude and altitude, and so on—which continue to surround the whole attempt at calibration based on the sequoia and bristlecone pine dates from the south-western United States. One part of my thesis here is relevant to McKerrell's case (1972, especially 293–7, and below) for a reconsidered calibration based on Egyptian historical dates. Since he has not yet turned his attention to these Aegean Late Bronze Age dates as a group, one can apply his methods to this material without incurring the charge of circular argument. In fact his 'historical correction curve', when applied to most of the significant dates in table 1—the last two from Pylos, the three from Mycenae, the one from Knossos LM IIIA, the two from Thera, the earlier of the two from Ayia Irini—gives results that are very satisfactory in historical terms. This seems to me to give independent evidence of the basic soundness of his conclusions, at least in respect of the second half of the second millennium BC.

I am aware that this paper could be open to the charge of a certain kind of play-acting, in that it is arguing from C-14 dates in an epoch for which we do not really need C-14 dates, since there is a historically derived chronology which, by its nature, is likely to be more accurate. But I think that there are genuine lessons here. If the relative proximity of documentary sources can save us from the pitfalls into which such evidence as the two Palaikastro dates would, on its own, lead us, then what errors, comparable or worse, may we not have fallen into in their absence? It is excusable to fall into ditches on a dark night, but it helps if one does not blindfold oneself into the bargain; and that is what we are doing if we ignore the evidence from contexts where there is some kind of independent check. But a part of the case that I have been presenting here—and I think it is the more substantial part—is wholly independent of questions of calibration. It is really no more than an appeal to the standards embodied in the concept of scholarly rigour. These standards do not of course hold all the answers; one must experiment and speculate. But they do require us, for example, to try to include all the relevant evidence; and above all, to try to ensure that we compare like evidence with like. I hope that we who are archaeologists will at least not turn our backs on them.

Correction Procedures for C-14 Dates

H. MCKERRELL

There are two quite separate criteria to satisfy before accepting as valid the corrections to C-14 dates which have been indicated for some years now by the bristlecone pine calibration. Firstly the correction figures have to be based upon all the available tree-ring data and derived in a manner that is mathematically sound, and secondly the correction figures have to produce accurate results on C-14 dates from archaeological test samples of known historical date, these covering as wide a period as possible. Neither of these basic prerequisites has yet been fully met. Thus the two-fold purpose of this paper is to bring together, and to compare with an independently based procedure, the various correction curves or tables that have been published up to Spring 1974, as well as to detail the correction results on reliable, historically dated Egyptian, Helladic and Minoan test samples from 3100 BC.

The nomenclature followed is strictly that adopted by the primary dating journal *Radiocarbon*, all C-14 dates quoted thus relate to the 5568 year half-life and the standard AD/BC system.

Deriving Correction Figures

From the outset one is faced with a major difficulty, which is that about half (nearly three hundred) of the available bristlecone pine dates, those from the La Jolla laboratory, have never been numerically published. However, they have been graphically published by Libby (1970), Berger (1970) and Suess (1970), but make use of average rather than individual standard deviations for each point on the graph. Even so the mean figures can be read with an accuracy of ten or twenty years, and have been incorporated in various correction tables such as those of Damon *et al.* (1972), Ralph *et al.* (1973) and Switsur (1973). The other three hundred recently determined tree-ring C-14 dates are those from the Arizona and Pennsylvania laboratories, published in the 1969 Nobel conference (Michael and Ralph 1970; Damon *et al.* 1970) and the 1972 New Zealand C-14 conference proceedings (Damon *et al.* 1972; Michael and Ralph 1972).

In general, agreement between all three laboratories is quite satisfactory, but it is not easy to find examples of comparisons on material of exactly the same dendrochronological age. Thus the first stage of this present exercise was to make an assessment of the data from the three laboratories, in terms of comparing the La Jolla results with either (or both) the Arizona and Pennsylvania results. This was done by taking all the results from each laboratory over periods of one hundred years at about 500-year intervals back to 7000 years BP in dendrochronological years. The average number of dates in each hundred years is about six for both La Jolla and Arizona plus Pennsylvania. It is obviously desirable to make the comparison with as many dates as possible and for these to be extracted from the minimum dendrochronological span. The hundred-year interval seems to be the minimum providing a satisfactory number of results. No 'bad' dates were excluded from the comparison and the only preliminary assessment, as to which century to examine, was to try and use a period with data spread evenly (in tree-ring years) throughout the century of choice. Table 1 lists all the data used. [See pp. 71–100 for all figures and tables.]

The comparison procedure used was the standard statistical *t*-test, which has the advantage of simplicity and is quite accurately appropriate to the situation under consideration. All statistical tests are to an extent approximations for actual as opposed to ideal data and this is certainly the case here. However, calculation demonstrates that the errors involved are insignificant, even for the most extreme cases found in the tree-ring data, and the use of the *t*-test is thus preferred here to more complex but less well-known routines. The method compares the mean date, over each century, for La Jolla and Arizona plus Pennsylvania, and allows one to put a percentage probability upon the likelihood of both sets of data being fully compatible. For such work the usual statistical line is drawn at the 5 per cent level: that is to say, only if the *t*-test probability falls below 5 per cent are we justified in stating positively that both sets of results probably differ significantly. The minimum probability recorded in table 1 is much greater at 20 per cent, and we may conclude that all the data compared is in good agreement throughout the complete period back to 7000 BP. Only a few comparisons were possible for Arizona and Pennsylvania results separately (using the same hundred-year periods) but again agreement is fully satisfactory. This is not to say that some individual results may not be seriously divergent. Such occurrences are a normal hazard of any laboratory activity and do seem to occur occasionally in the data under consideration, but here they are a minority feature in what may reasonably be regarded as an excellently internally consistent series. We may thus proceed to an examination of the data as a whole, with a view to establishing some simplified relationship between the dendrochronological age and the C-14 age.

At this point we may make brief mention of a basic methodological problem in establishing any such relationship. Essentially, what is ultimately required is some method of converting experimental C-14 dates into calendrical (dendrochronological) dates, i.e. dendrochronological age to be expressed as a function of C-14 age by a relationship of the form: dendrochronological age=*f*(C-14 age). Any averaging of data has essentially to be carried out on the variable parameter, which is the C-14 age. For example, if we are considering which C-14 date to apply to which dendrochronological date, and if we have ten experimental C-14 dates on the same tree-ring sample, then this will allow of calculation of a mean C-14 date accurately applicable to that particular ring date. But any attempt to select a C-14 date (a single year) and put to it an average dendrochronological date is clearly a poor and less accurate use of all the data. Although this is a somewhat extreme example of the actual data as we have it, the same situation does obtain fully in practice and it is unfortunately necessary to generate a relationship of the form: C-14 age=*f*(dendro. age), i.e. we express the C-14 age as a function of the dendrochronological age. There are a number of ways of resolving this difficulty and providing a relationship yielding dendrochronological dates as a function of C-14 date: the most straightforward, and that adopted here, is to express the relationship graphically. By interpolation one can read off the resulting curve, for any chosen C-14 age, the equivalent dendrochronological age.

Date Averaging

By averaging the experimental dates contained within each fifty tree-ring years, we can derive effectively a smoothed version of the tree-ring curve. Obviously it is desirable to keep the number of tree-ring years to a suitable minimum and clearly the fifty years mentioned would seem preferable to one hundred. But would twenty-five years be better still? The solution is not quite as subjective as it appears, because it is clearly necessary to include within the tree-ring bracket sufficient C-14 dates to make the average meaningful. With about 600 dates spread (somewhat unevenly) over 6000 years, we average about ten per century or five per fifty years. Clearly then there would be little point in reducing the bracket to twenty-five years. The standard error of the mean of ten dates within a hundred-year bracket will be the standard deviation divided by $10^{1/2}$. For the same standard deviation and with five dates we would divide the standard deviation by $5^{1/2}$. Thus, although we double the number of dates, the use of ten as opposed to five only decreases the standard error of the mean by about 1·4 times. The procedure adopted was to take the dates contained within successive fifty-year tree-ring brackets and to calculate the population mean C-14 dates applicable at the mid-point of each tree-ring bracket. However, for some periods there are insufficient dates available and the bracket was necessarily widened to one hundred years. Whenever possible, at least four dates were averaged.

The resulting plot of C-14 date means against dendrochronological date is illustrated in figures 1 to 4. The upper and lower limits are plotted for plus and minus one standard error of each mean which corresponds to the usual 68 per cent confidence limits encountered with a single C-14 date standard deviation. It is apparent that the additional smoothing which could be introduced by the technique of a moving fifty- or one-hundred-year average would not be significant, and the basic successive fifty- or one-hundred-year average has therefore been used throughout. Inspection of the relevant plots shows that many of the minor fluctuations of the La Jolla/Suess curve are smoothed out, although the more positive fluctuations are maintained. A particular difference is that the feature of derived alternative calendrical dates corresponding to certain C-14 dates is replaced largely, though not completely, by date *ranges* usually narrower than the overall span from the alternative figures. The merit of the data as plotted is that it is based upon standard mathematical procedures strictly applicable to the type of data involved, and any one mean as plotted will relate to the true mean, for the same tree-ring period, through the standard error as detailed. For certain periods, where only a few determinations are available, the excess scatter generated by a larger than usual standard error may be somewhat misleading if it alone were to be interpreted as the cause of short-term fluctuations. Where this has occurred, and where the mean itself indicates no such fluctuation, the upper and lower limits have been extrapolated from more reliable data on either side of the point of interest. The results obtained overall, and the curves generated, should therefore be correct and significant within the limits mentioned. The fluctuations that still remain do thus have a sound statistical basis.

Polynomial Regression Analysis

This standard technique for handling many data was carried out by computer, and polynomials of up to twenty orders were used. Essentially, the higher the order the better the fit to such a fluctuating curve as that here involved, also the lower will be the standard error of the mean. However, above a tenth-order polynomial no really significant improvement in fit or standard error was to be found, and for convenience the results are detailed up to this order only. Table 2 lists derived results showing the fit obtained, and figure 5 illustrates a sixth-order curve running through the individual data points. The two methods of deriving the results, described earlier, yield different data according to which parameter (C-14 date or dendrochronological date) is treated as the variable. As shown in table 3, the use of regression analysis equations directly to convert C-14 dates to dendrochronological dates results in errors of up to sixty years compared with the more rigorous procedure of deriving C-14 dates from dendrochronological dates.

In general, the agreement between the fifty-year averaging procedure and polynomial regression is quite good. Thus, from the curves in figures 1–5, a number of C-14 dates were selected which would be predicted as likely to afford both good and bad agreement. The good results are listed in table 4 and, for the higher order polynomials, both methods differ by less than forty years. This is quite good, but for other dates, chosen at regions of significant fluctuation, the agreement is much worse: table 5 indicates that differences of up to three centuries can occur. Such severe divergence would seem to be an inevitable consequence of polynomial regression analysis when dealing with such variable data. The fluctuations in the fifty-year averaging curves, detailed in figures 1–4, do seem soundly based and it is difficult to believe that their removal by regression smoothing is valid. Essentially, the technique does not seem able to cope with the severe curve fluctuations involved, and significantly over-smoothes the data in some regions. For this reason such curve-fitting procedures were not considered further for the tree-ring data.

Short-Term Fluctuations

The curves plotted from the fifty-year averaging procedure (figures 1–4) quite clearly retain a number of wiggles, which imply that such short-term fluctuations are real and do exist. The tree-ring curve fluctuations based upon the La Jolla results alone do seem less soundly certain, being largely a forecast of the situation that *might* be found, eventually, and with more data (Suess 1970). Many of these fluctuations, however, are confirmed by the averaging procedure described, and it is difficult to reconcile this with conclusions elsewhere that such detailed structure in the La Jolla curve is not necessarily valid. One important refutation of some of the La Jolla tree-ring curve wiggles is based upon the archaeological evidence from a number of Late Neolithic enclosures (Burleigh *et al.* 1972)—it is thus important here to consider briefly the source of these apparent contradictions.

The first approach to the data was to consider the Mount Pleasant sequences observed both stratigraphically and by the laboratory determined dates. Table 6 lists all the dates and their provenances. Correct stress was placed by the authors involved upon the elegant correlation thus observed and figure 6 (left) illustrates interpretation of the evidence. However, the vital conclusion that this order would be removed after correcting the dates by the La Jolla tree-ring curve does not seem to be borne out. Thus the right half of figure 6 compares the same date series after tree-ring correction, and it is apparent that a very similar sequence ordering is obtained. The overall relative lengths of some of these periods is certainly changed, this being one primary result of the correction method, but essentially a similar picture emerges and one may even point to the overlap observed for both series between phases 1 and 2, 3 and 4, and 5 and 6 (present nomenclature *re* figure 6) and the (slight) gaps between phases 2 and 3 and 4 and 5. So although some changes do appear after tree-ring correction, it is not really possible to agree that this is at all contradictory to the archaeological evidence. Or, viewed another way, the site stratigraphy does not seem to invalidate the corrections derived by the La Jolla tree-ring curve.

The important pottery date sequences were also examined. Tables 7 to 12 list the basic data in chronological order and for each ceramic phase described, and table 13 lists the dates for the three Continental beakers used comparatively. As with the Mount Pleasant sequences, when these are plotted (figure 7) they illustrate the succession and overlap discussed in detail in the original article. The clear ordering observed in the basic laboratory C-14 dates is again apparent in the tree-ring corrected data. As before, because of the shape of the correction curve at the period of interest, most of the phases are chronologically longer after correction, but no order reversals occur and the overall impression generated is very similar whichever sequence is examined. Thus the Continental beakers overlap with the grooved ware and the appearance of the few earliest beaker sherds. And the succession, Mid-Neolithic plain bowls, grooved ware, grooved ware and beakers, late beakers and finally collared food-vessels, is apparent in both sets of data. Again it is difficult to see the ceramic sequencing as providing serious contradiction of the tree-ring curve minor fluctuations.

All in all the henge monument data could be viewed as not totally supporting the arguments set forth by the authors of the original article. Certainly their general appraisal of the confusion likely to be caused in this region of the tree-ring curve seems totally correct, but it may well be that with sufficient high-accuracy results such conclusions are not always valid. Certainly, as far as the present situation is concerned, the wiggles from even the La Jolla curve are not strictly invalidated by such archaeological evidence. The fifty-year averaging curves which maintain such short-term fluctuations do thus seem realistic.

Comparison of Correction Curves

Appendix IA lists comparative tree-ring dates from various correction curves or tables at fifty-year intervals of C-14 date. Essentially the only curve-fitting analysis listed is that by Wendland and Donley (1971), although the figures recommended by Switsur (1973) were derived partially from the work of Damon *et al.* (1972) who also used a regression curve-fitting procedure. The problems listed above, in terms of difficulties relating to such techniques, mean that the Wendland and Donley curve does severely over-smooth the data. Switsur's figures, the mean of the 1972 Arizona and Pennsylvania corrections, which provided the best data available at the time of publication, would also be expected to somewhat over-smooth areas of significant short-term fluctuation (following from the Arizona results), and figure 8 illustrates this point. The original La Jolla tree-ring curve (Suess 1970) clearly produces what appear to be over-excessive ranges of correction relative to more recent data, a feature following from the extensive incorporation of wiggles in the basic curve. However the recent MASCA assessment (Han *et al.* 1973) yields very close agreement to the fifty-year averaging figures detailed above, as figure 9 illustrates.

Essentially, then, the fifty-year averaging procedure produces similar results to the MASCA nine cell treatment, and fairly good agreement with Switsur's data. All three series could be used with confidence for correction purposes, though preference is here expressed for the MASCA or fifty-year averaging data. Unlike the MASCA procedure, which corrects to best mean figures, the fifty-year averaging provides a correction band at all points. Even so there is still a 32 per cent probability that the true dendrochronological equivalent to any C-14 date will be outwith the ranges quoted. Appendix IB provides tree-ring corrected figures at intervals of ten years for C-14 dates over the range 4500 BC–1800 AD, and effectively satisfies the first of our criteria for acceptable correction data; i.e. results based upon all the available tree-ring C-14 dates with analysis effected by a mathematically appropriate procedure yielding close agreement to other independently and soundly derived figures. The correction figures obtained may now be utilised in assessing the accuracy of correction using historically dated samples, largely ancient Egyptian, of certain provenance.

Egyptian and Other Historical Data

The BM/UCLA series of Egyptian historical dates (Berger 1970; Edwards 1970) is the most useful data for comparison with the tree-ring curves. Most of the material in this series has been obtained and dated during the last few years and represents the best of its kind. The dates published in the BA/Royal Society proceedings of 1970 (Berger 1970; Edwards 1970) have been modified in a few cases, and I am indebted to colleagues at the British Museum for valuable pre-publication advice here. The use of Egyptian dated material in checking the validity of C-14 dates has been recognised from the first with Libby's work, but invariably the samples used have not been at all ideal and a good summary, perhaps somewhat over-critical, is provided in the Nobel proceedings (Säve-Söderbergh and Olsson 1970). The particular virtue of the BM/UCLA series is that it was obtained primarily with C-14 dating in mind. Samples are mostly of excellent provenance and generally do not suffer from the problems caused by using massive timbers which could be many years older than the events of interest. Even so, some of the samples in the lists are not quite as impeccable as is here necessary, and where material has been excluded from consideration it is listed, with reasons, in table 14. Tables 15 and 16 detail all the data used in the comparison and include the overall historical dating accuracy associated with the samples. The basic chronological scheme followed is that of Hayes as detailed in the Cambridge Ancient History series (Hayes *et al.* 1962). There is additional error from the excavation itself and wherever possible a contribution to the overall accuracy has been included from this source. The dates thus listed should be historically correct within the limits in tables 15 and 16. Essentially, the figures obtained are either identical with, or extremely close to, those listed by Edwards (1970). Where there is any divergence at all the reason is listed in table 17. The historical dates for the UCLA samples have been carefully maintained as the same as those for the BM—this was not always done in the original publication (Berger 1970). Also, some of the published UCLA C-14 dates were confused between different tables and text in their original publication—the figures listed in table 16 are believed to be now correct. Any doubtful samples are, as mentioned earlier, excluded and listed in table 14.

Inter-Laboratory Comparison for Egyptian Dates

The first point to consider is the reliability of the Egyptian C-14 dates, firstly in terms of comparison between the two laboratories and secondly in terms of statistical breakdown for all results of the same historical date. We have examined the inter-laboratory aspect elsewhere some years ago (McKerrell 1971), but as the figures then used have altered slightly in some cases (BM improvements, recalculation etc.) this is repeated here. Table 18 lists the comparison pairs: Edwards gave twenty-four pairs and as one of these is excluded by table 14 (being apparently a printing or measurement error) we are left with twenty-three direct comparisons. As far as possible throughout this whole exercise comparison and examination has been made in terms of the factor most widely used by archaeologists, the standard deviation. This was therefore applied first of all to the inter-laboratory comparison, and the date differences are expressed in table 18 in terms of the numbers of the standard deviations that they represent, the difference SD being calculated from the two pairs' SDs according to $SD=(SD_1^2+SD_2^2)^{1/2}$. Figure 10 is a plot of this data. The question to be asked is whether the series of results has given the SD differences that would be predicted for statistically normal data. Figure 10 therefore groups the results over each unit of standard deviation, positive and negative, and the resulting frequencies may be compared with those predicted both theoretically and from a series of twenty-three experimental comparisons generated by radioactive decay and therefore entirely normal. In all, the C-14 results give thirteen positive SDs and ten negative SDs: this is, by chance, exactly the same as the breakdown for this particular experimental comparison and quite clearly there is no systematic bias in terms of one laboratory consistently producing older or younger dates than the other. So far, so good. But what of the frequencies? Here again it would be difficult to see anything but a normal (statistical) situation. There are no SDs above 2·6 (the maximum figure is −2·6) and the percentage falling between zero and one SD is 52 per cent, between one and two SDs 39 per cent, and above two SDs 9 per cent: these compare with the theoretical 68, 27 and 5 per cent. Table 19 lists all the frequencies and calculated percentages; theoretical, C-14 and experimental.

It is clear that the inter-laboratory comparison shows a situation very close to the normal, as even the most casual inspection of figure 10 would suggest. More sophisticated tests are available to add to such conclusions but these are not detailed in view of the clarity evident in the foregoing. However, I am greatly indebted to the Department of Statistics at Edinburgh University for advice and facilities in this matter. The Cramer von Mises test, for example, is particularly appropriate and indicates no significant probability (less than one per cent) that a non-normal relationship exists in the two sets of data. These conclusions could scarcely be more at odds with a recently published comparison by Clark and Renfrew (1973) on, apparently, the same figures. It is in fact difficult to believe that the authors of this work are using the correct data—either this or serious arithmetical errors obtain—but in the absence of details of the figures actually used it is not profitable to pursue further speculation. However, particularly obvious discrepancies—such as inclusion of UCLA-1398, a result that is eight standard errors deviant from the mean of five other contemporaneous samples (see table 14) and, apparently, a completely non-critical acceptance of provenance, printing errors and differing historical date assessments between the BM and UCLA on identical samples—all combine to cast severe doubts on the conclusions reached. Such an unfortunately casual approach does no justice to the subject itself and even less to the excellent laboratory work involved.

One additional test, however, is again the standard *t*-test, which can be used to compare groups of C-14 dates of the same historical date from each laboratory, and to provide a positive statement that the two sets of dates are, or are not, compatible. The procedure is as rigorous with small numbers of comparison dates as with many dates and only one detail needs to be considered. Thus, if we are comparing two groups of dates, each group being entirely normal with perhaps different variances (SD), then strictly standard normal comparison tests apply. The implicit assumption is that we are comparing data from only one, or at most two, universes. Our C-14 dates may not be quite so impeccable, although if each date in the respective groups is of identical variance (as determined by the laboratory SD) there is no problem: we are comparing two precise universes with perhaps different means but identical variances. But if one or both groups contain dates of different variances then clearly a single group can contain data from more than just one universe. Even if the mean of each of these individual universes be the same, the variance differences will put the data into different universes, and to lump all the results in this group together, to calculate a mean and overall standard deviation, could at first glance be incorrect. To anticipate somewhat we may state that the error involved is trivial (about three per cent) but, since the point is important with the kind of data under consideration, it is worth demonstrating.

For example, consider two groups of data with the same mean, group 1 having one hundred points and an SD of fifty years, group 2 also having one hundred points but with an SD of one hundred years. If we average all these two hundred dates, we would obtain a standard deviation of very nearly seventy years for the mean, the mean being the same as both initial groups. But whilst 68 per cent of all the two hundred dates would now fall within zero to one SD, we would find that 24 rather than 27 per cent would fall between one and two SDs, and eight rather than five per cent above two SDs. Thus although the procedure is not strictly rigorous because we are averaging data from different universes, we may note that even with one SD being twice as great as the other (chosen to be in excess of anything encountered here) there are only three dates in a hundred that are misplaced, and these only from the bracket 1 to 2 into 2 to 3 SDs. Clearly there is little error involved in averaging, say, up to six dates or so, some perhaps with different standard deviations, and taking the calculated SD and standard error of the mean to be applicable to the group mean.

This was the procedure adopted for the inter-laboratory comparison *t*-tests, and table 20 lists all the data. All the dates were used that covered the minimum of various historical spans and the aim was to obtain the maximum number of BM and UCLA results with, at the same time, the minimum historical spread: the relative dating of many of the samples involved is often considerably better than their absolute values. The dates were also chosen to cover the complete historical period between 600 and 3000 BC. And with no resulting probabilities below five per cent for the differences between BM and UCLA being due to chance, it is clear that the comparison indicates very positively that there is no significant variability between laboratories above that expected for normal data.

Polynomial Regression Analysis

Two methods of comparison with the tree-ring data were employed. Firstly, all the Egyptian dates listed in tables 15 and 16 were used in a polynomial regression procedure to provide a simple graphical relationship (figure 11) for comparison with the tree-ring curves obtained earlier. The upper and lower limits were obtained using the overall historical limits listed, and low order polynomials were entirely adequate: figure 11 is drawn using a fourth-order regression curve. The points plotted are the means of samples of the same historical date and, as may be observed, the overall fit to the curve is very good. It was not expected that anything like the complexity of the tree-ring curves could possibly be obtained, and the regression curve fit seems a reasonable method of using all the data. Within the limits drawn it should be possible, using figure 11, to convert an experimental C-14 date to an absolute historical date, and this in turn should compare with the tree-ring date derivable from figures 1–4. The agreement however is in fact rather poor back to c. 2300 BC, though from then to 3000 BC it is excellent. In particular, the data around 600 BC, 1200 BC and 1900 BC do seem seriously off the tree-ring plot and the two curves, as drawn, are virtually incompatible for nearly all the regions for which we have adequate dates back to c. 2100 BC in calendrical years. Appendix IC lists the corresponding historical dates for standard C-14 dates at intervals of ten years.

t-test Comparison with Tree-Ring Data

The second method of comparison leads to similar conclusions. This involves a statistical *t*-test procedure and compares the means of C-14 dates for samples of the same historical age with means of tree-ring C-14 dates for samples with the same calendrical date range. Use of this technique has been restricted to historical dates which have, as a minimum, three associated experimental C-14 dates. As mentioned earlier, the *t*-test allows of a probability statement that both sets of data are, or are not, significantly different; only if the probability falls below five per cent are we justified in concluding that the two compared sets are in fact significantly different. The procedure for dealing with the historical dates was to compare these, firstly, with all the tree-ring dates covering the complete plus or minus date range for the historical material of interest. However, since this range may be somewhat large (up to, for example, two hundred years) and since the historical material is not necessarily spread evenly over this range (in fact in all probability it falls at a very narrow, but unknown, bracket inside the overall range), an ideal comparison cannot be made. But what can be done is to split the date range into two or three even parts (e.g. if there are sufficient tree-ring dates, a lower third, a mid third, and an upper third) and then carry out additional *t*-tests on these two or three portions, comparing each with the historical results. Only if each probability is broadly similar (for these and the overall range) may we then draw any conclusions as to significance. This is an extremely demanding procedure but it was felt to be necessary if really positive conclusions were to be made. Tables 21–25 list the data for the periods mentioned above as providing apparently poor agreement, and tables 26–28 for areas of expected good agreement.

There are four Egyptian results dated historically to 600±70 BC. t-test comparisons show that for tree-ring dates falling within the dendrochronological range 670 to 530 BC there is a 98·5 per cent probability that the two sets of data are significantly different. Similarly, for the tree-ring periods 600 to 530 BC and 670 to 600 BC the probabilities are 99 per cent and 97 per cent. We may therefore conclude from this that the Egyptian dates around 600 BC do differ significantly from the tree-ring dates around 600 BC. There seems no good reason to suspect that the Egyptian dating is invalid; all the samples are positively from the Saite period in the 26th Dynasty, which lasted from about 664 to 525 BC. Nor is there any reason to suggest that the dated samples are significantly older than their burial date; none were taken from massive timbers and indeed one of them was reed material. Taken as a whole they do seem a consistent group. Similarly, the tree-ring dates from La Jolla, when compared with those from Arizona and Pennsylvania combined, agree extremely well, with about an 80 per cent probability that both sets of data are compatible, over the range 670 to 530 BC. All the data are in table 21.

Eight of the Egyptian results are dated historically to around 1200 BC, and four of these, put at 1170±30 BC, have been examined separately from the total eight. Within this range, 1200–1140 BC, there are in fact only two tree-ring dates with which to compare but widening the limits by just ten years in either direction increases the comparative tree-ring sample to six, and these are the comparison results used. Thus table 22, following t-test examination, shows that there is a better than 99 per cent probability that the two sets of data are incompatible. It was not considered profitable to attempt to split the few tree-ring dates into groups but, to provide an extra test at this period, all the eight samples falling within the range 1254 to 1140 BC were examined (table 23). t-tests show that there is a 98·5 per cent probability that this data and all tree-ring results from the same period are incompatible. Similarly, tree-ring dates between 1200 and 1140 BC, and 1254 to 1200 BC, show respective probabilities of 98 per cent and 96 per cent for incompatibility. It is then highly likely that the Egyptian dates and the tree-ring dates do differ significantly at around 1200 BC. By contrast, the tree-ring data from La Jolla when compared with that from Arizona and Pennsylvania combined, over the same date range, shows a high degree of compatibility at greater than the 90 per cent level. It is again difficult to believe that, at this period, there is any serious error in the Egyptian dating outwith the limits listed. The samples involved relate to funerary monuments of the 19th and 20th Dynasties, including that of Rameses II and prominent prophets and priests. All the samples were collected by Martin in the field, and were positively archaeologically sealed to the phases involved. Moreover, half the samples are reed and the other half branches, and all were embedded in mud-brick courses contemporaneous with the constructions of interest. There would seem, therefore, no reason at all to suspect that the samples dated can be at all older than the activity they represent; reed and tree branches would be generally considered excellent materials. As with the samples around 600 BC, we seem to have a positive discrepancy between historical and tree-ring dating.

Three C-14 dates relate to the death of Sesostris II in 1880±25 BC. For such a narrow range only four dendrochronological samples have been dated, though table 24 shows that the *t*-test examinations put the probability at 98·5 per cent. However, to allow further examination the tree-ring comparison was extended to 1880±50 BC, and *t*-test results show that all the tree-ring samples in this period have a 99·9 per cent probability of being incompatible with the Egyptian dates. Similarly, tree-ring samples over the ranges 1889 to 1830 BC and 1930 to 1880 BC have probabilities of greater than 99 per cent. As noted on other occasions, however, the tree-ring data from La Jolla and Pennsylvania and Arizona combined are highly compatible (table 24). With the recorded heliacal rising of Sirius in 1872 BC, during the seventh year of the reign of Sesostris III, we have one of the fixed points of ancient Egyptian chronology. Depending upon where the sighting was made there is a possible error of at most about twenty years (generally considered as excessive and occasioned only by the unlikely possibility that sightings were made at Thebes). The death of Sesostris II (who preceded Sesostris III) would thus be put at about 1880 BC with an overall, and perhaps excessive, accuracy of ±25 years. This is probably the most reliable date of the series that we are here considering. Moreover, the samples relating to the death of Sesostris II were all obtained directly by Martin from the funerary monument itself, and as the material collected was reed there can be no question of non-contemporaneity with the date of burial. All in all these three dates are probably the most reliable of the whole Egyptian series and, as described, are quite incompatible with any related tree-ring dates.

Six dates were obtained on early material from the tomb of Intef. As described in table 14, one date has been excluded from consideration here. This is UCLA-1398, which is 309 years younger than the mean of the other five; the standard error of the mean of these five is thirty-nine years so that the deviation is equivalent to very nearly eight standard errors. There can be no question of this occurring by statistical chance alone and some laboratory error or mis-print seems indicated. Table 25 lists the data comparing the results with the tree-ring data, using *t*-test statistics, and in no case can we see less than a 99 per cent probability of incompatibility. However, the historical dating of these samples requires some explanation. Edwards (1970) puts Intef's death in the late 11th Dynasty or some time in the early 12th Dynasty, i.e. c. 2000 BC, but Settgast, the excavator, states in fact that the tomb relates to some time between the late 11th Dynasty and the *late* 12th Dynasty, i.e. 1920±100 BC (Arnold and Settgast 1965). It is apparent that this wider estimate does cover the c. 2000 BC date, and we adopt here the figure of 1920±100 BC. Table 25 indicates, with greater than 99 per cent probability, that the five Egyptian dates are quite incompatible with tree-ring dates for any of the periods 1920–1820 BC, 1970–1870 BC or 2020–1920 BC. As on previous occasions, there is adequate compatibility between the La Jolla tree-ring dates and the Pennsylvania and Arizona dates combined over the period 2020 to 1820 BC. Finally, a *t*-test calculation was completed on the basis of the c. 2000 BC dating, put at 2000±25 BC. This showed a 99·5 per cent probability that the two sets of data were again incompatible, and we may conclude that, no matter which dating is accepted for General Intef's tomb, the archaeological dates are quite incompatible with the tree-ring dates. There is no reason, either, to suspect that we are dealing with samples much older than Intef's burial: the average C-14 date of 1689 BC compares well with the Sesostris II burial average of 1641 BC, and suggests that the Intef samples can only be a little older than the reed from the precisely dated tomb of Sesostris II. Since this latter is fixed at 1880±25 BC, it is apparent that the 1920±100 BC date for Intef's tomb fits extremely well. To suggest that the Intef dates refer in fact to a date older than the 2020 BC likely maximum would not appear at all realistic. Further corroboration is provided by a consideration of all the Egyptian dates in this region, and table 29 lists the various means, etc., while in figure 12 the C-14 dates are plotted against the historical dates. It may be observed that the dates fall into a simple consistent linear pattern, and to contemplate the Intef dates being significantly older than 1920 BC is clearly at odds with all the dating evidence.

In summary, then, it is clear that for the four periods after 2000 BC, where we have groups of three or more historical dates to deal with, there is total incompatibility with the tree-ring curve. That is to say, the C-14 dates obtained for these historically dated samples would all be made too old if we were to rely upon the tree-ring corrections alone. The general impression of the regression curve for the Egyptian data (figure 11) is clearly that of a fairly smooth line not subject to any severe changes in slope. And, although the data back to 2000 BC are not by any means as complete as we would wish for, it is probable that the extrapolated portions, covering areas for which adequate data are not yet available, are meaningful within the limits drawn.

However, for the historical period between 3100 and 2300 BC the agreement with the tree-ring curve is excellent. Tables 26–28 list the comparative *t*-test results, and for all the periods where we have sufficient data, namely 2350–2335 BC, 2650 BC and 3025–3000 BC, there is a high degree of compatibility with the tree-ring data. Moreover, for many samples with just two dates (and therefore deliberately excluded from the more rigorous mathematical procedures) inspection of figure 11 shows that all the data from this period fall quite well on the historical curve and hence also the tree-ring curve.

Helladic and Minoan Data

The Egyptian historical series of C-14 dates is far and away the best material for comparison with the tree-ring data, but dates from two Aegean sites, Lerna (Kohler and Ralph 1966) and Myrtos (Warren 1972), are also from contexts of good reliability and cover the historical period of 1800 to 2200 BC that, as we have already seen, involves severe discrepancies between tree-ring and Egyptian samples. The Aegean historical chronology is calibrated by the Egyptian calendar and is generally considered reliable over this particular time span. The material thus affords a comparative check on the discrepancies already noted.

Seven Myrtos dates refer to the destruction levels of the palace site, and date, archaeologically, to the end of Early Minoan IIB. A date range of 2400–2170 BC is accepted in the Myrtos publication, and current discussion in the literature would seem concerned only to adjust the end of EM IIB by a few decades at most. It seems unlikely that a range of 2200–2150 BC would be far off the mark for the destruction levels, and this would be the historical date for all the Myrtos samples listed in table 30.

The Lerna samples, C-14 and historical dates are listed in table 31. The excavator uses an internal site level dating scheme, the absolute dates to each level being derived from the conventional Helladic chronology, itself in turn calibrated by the Egyptian calendar. The site descriptions, e.g. Late Lerna III, Early Lerna IV, etc., have thus been applied to the absolute dates for each level and the resulting sample historical dates are those listed in table 31. An absolute accuracy of about ± 50 to 75 years is probable.

Two of the Lerna samples dating historically to 2200 BC have been combined with the 2170 BC samples from Myrtos, and all nine were used in a *t*-test comparison with the corresponding dendrochronological C-14 dates. There are twelve tree-ring dates covering the period of the historical samples 2250–2150 BC, and table 32 lists all the data. The probability of the Myrtos and Lerna dates being incompatible with the tree-ring results is about 99·9 per cent; in contrast the comparison of tree-ring results from Arizona, and La Jolla and Pennsylvania combined, indicates entirely adequate compatibility.

An attempt by Harding (1973) to reconcile the historical dating involved with the Myrtos corrected C-14 dates seems clearly at odds with the foregoing, and it would be worthwhile to briefly examine this discrepancy. From the outset there is confusion as to the historical dating he uses; Warren clearly states (1972, 343–5) that the samples involved relate to the final palace destruction at the *end* of EM II, not generally EM IIB as implied by Harding—EM IIB covers perhaps two-and-a-half centuries so this is no mean assumption. Moreover most of the samples were twigs or fallen roof timbers and in two cases possibly the remains of a ladder. None came from massive timbers that could be suspected of being some considerable age either at the time of initial use or site destruction (such as from Pylos; Kohler and Ralph 1961). Nor, with the usual ceramic transition from EM IIA to EM IIB, well represented at Myrtos, and a substantial accumulation of EM IIB material present, is there any reason to suspect that the samples do not relate to the end of EM II, c. 2170 BC. Figure 13 details the C-14 dates involved and their tree-ring corrected date ranges at limits of one standard deviation, using appendix IB. Of the seven samples only two are acceptably corrected to 2170 BC, and these in each case at the extreme lower limit of the corrected date range. Only three such corrected date ranges fell within EM IIB and, even allowing for the fact that Harding is referring to corrections based upon the La Jolla curve (Warren 1972, 344), it is patently inaccurate to describe 'almost all' the corrected results as falling within the EM IIB period, the relevant historical date for comparison being anyway the end of EM IIB. And to use a thermoluminescence dating of 2470±350 BC as specific confirmation of such conclusions is to totally ignore the TL accuracy stressed by both Fleming (quoted in Warren 1972, 343) and Warren (1972, 343)—the TL date range of about 2800–2100 BC in fact allows almost any likely option. So it is hard to see that Harding's conclusions seriously dilute the near certainty of the Myrtos C-14 dates diverging from the relevant historical date, after tree-ring correction, as noted above.

The remaining Lerna dates have been combined with the appropriate Egyptian dates for *t*-test comparisons around 1900 BC, 2000 BC and 2100 BC (table 33). It is apparent from inspection of the results that there is a 99 per cent, or greater, probability that the historical dates are quite incompatible with the tree-ring data.

The point to be made from these conclusions is essentially that inclusion of the Helladic and Minoan dates does not improve in any way the compatibility of historical and tree-ring C-14 dates around 2000 BC. But it would not be appropriate to put the emphasis any stronger than this because, although Lerna and Myrtos provide the best Aegean C-14 dates of this period, the data do not really compare in precision with the Egyptian material. Nonetheless they do positively confirm the pattern derived from the Egyptian series, and a serious deviation from the tree-ring data is indicated across the three related historical chronologies, and across a variety of excavations and dating laboratories. This is summarised in figure 14, which compares the original La Jolla tree-ring curve and the fifty-year averaging curve with all the historical dates. In order to provide the maximum possible compatibility for overlap, figure 14 includes two results from this period excluded in table 33 (UCLA-900 and UCLA-1398, though not UCLA-1211 which is of very doubtful provenance) as well as two appropriate results from the Groningen laboratory. These latter are precise, modern dates and GrN-1177 compares directly with BM-335 (historically 2000±50 BC) whilst GrN-1178 is of the same provenance as UCLA-900, the alleged funerary boat of Sesostris III, c. 1850 BC. These dates are listed in table 34.

Out of these thirty-two historical dates around 2000 BC, thirty-one yield C-14 results older than the comparable tree-ring results. The single exception is from a group of six contemporaneous samples and this is the highly suspect result mentioned earlier that deviates from the mean of the other five by 309 years, or eight standard errors of the group mean. Tree-ring statistical comparisons between the three different laboratories involved show no reason to doubt the tree-ring C-14 measurement reliability. And the use of historical dates from Crete, Greece and Egypt, as well as dates from five different laboratories, yields a totally consistent pattern of older C-14 dates than would be predicted by the tree-ring data. We simply cannot accept that all these laboratories, or all the excavators involved, are, by extraordinary coincidence, producing the same errors, nor that the basic historical chronology is several centuries too young in a region where existing historical documents record astronomical sightings of high precision. There is a consistency and a pattern about the situation which indicates that a severe problem does exist: that in certain regions of time, bristlecone pine tree-ring corrections to archaeological C-14 data do not yield the correct calendrical dates.

The examination by Snodgrass of Late Bronze Age Aegean C-14 dates, detailed above, affords very similar conclusions.

The assessment by Clark and Renfrew (1973) of the agreement between ancient Egyptian historical dates and tree-ring corrected C-14 dates on the same BM/UCLA series, yielded the conclusion that the calibrated C-14 dates did not differ significantly from the historical dates over the entire period 3000–1800 BC (calendar years). Doubts have already been expressed above of the usefulness of this work, and such a conclusion is clearly entirely at odds with the evidence detailed for samples around 2000 BC. No amount of statistics or verbal rhetoric could conceivably reconcile the c. 2000 BC historical results with their tree-ring corrected figures. The main cause of this discrepancy is that the tree-ring C-14 dates around the 2000 BC tree-ring period, as used by Clark and Renfrew, are not the basic data published graphically by Suess (1970). They used instead data from an unpublished PhD thesis, and assumed that these are identical with Suess' published figures. In fact this is not so, and figure 15 illustrates the deviation of the various C-14 dates from Suess' curve around 2000 BC in tree-ring years. Suess' data fall closely and evenly about the curve as would be predicted, but Clark and Renfrew's data, whilst claimed to be those of Suess, are very clearly quite divergent from the basic results. Likewise the Egyptian, Minoan and Helladic data, as noted above, are systematically divergent from the tree-ring curve and, moreover, are divergent in the same direction as the data used by Clark and Renfrew. It is thus not surprising that these authors conclude there is no difficulty in reconciling *their* tree-ring data with the historical dates around 2000 BC. But these are not the tree-ring results used by Suess (1970), Libby (1970), Berger (1970), Ralph *et al.* (1973) or the present writer (McKerrell 1972). Clearly Suess' data are not compatible with the historical dates, and Clark and Renfrew's attempt at reconciliation is quite invalid around 2000 BC.

Comparative Correction Curves

Although appendix IB details the tree-ring dates corresponding to given standard C-14 dates, perhaps a more straightforward procedure, and one moreover that allows a personal assessment of the validity of any particular correction, is to use figures 16 to 18. These plot the actual *corrections* to be added to standard 5568 C-14 dates, and are based upon the fifty-year averaging procedure detailed earlier. The curves are drawn over plus and minus one standard error of the mean of each fifty-year tree-ring group, and there is thus a 68 per cent probability that the true correction lies within the curves as drawn. The approximate nature of the magnitude of any correction is thus clearly represented and the curves are a deliberate attempt to avoid placing any reliance upon a single point correction, as is normally the case. The corrections themselves are as subject to statistical variation as any single C-14 date and this has to be recognised in any conversion procedure used. As drawn, the correction curves take account of fluctuations in the basic tree-ring plot as well as the statistical parameter of one standard error.

In addition to this, figure 19 superimposes upon the tree-ring corrections the historical corrections from the evaluation of the Egyptian historical data, derived from the polynomial procedure detailed earlier. The regions of divergence and agreement are apparent from figure 19, and direct comparison of the magnitude of either type of correction is quite straightforward.

To point to the discrepancies after 2000 BC is one thing, to account for them quite another. There seems no obvious reason of a physical nature, and we have gone to some lengths to demonstrate that both the historical material and the tree-ring data are not suspect in themselves. Had the discrepancies been restricted to a few samples, or to just one period of time, the problem would be of much less consequence. But this is clearly not the case. As far as we can tell, from the best data at present available, there is a consistent, significant discrepancy throughout much of the historical period back to 2000 BC. Quite what to do with this conclusion is equally not too obvious. Certainly more accurate dates will only help but if, as we suspect will be the case, the present results are confirmed severe concern must be expressed over the ultimate accuracy of correcting C-14 dates. For even if we can correct dates against a much improved Egyptian historical curve, what guarantee is there that this is the best correction procedure for, say, Western European Early Bronze Age dates? Why this curve rather than the tree-ring curve, or indeed any other curve?

One detail that does require to be stressed, however, is the reason for wishing to correct archaeological dates in the first place. Essentially, the basic C-14 data are best used directly for many comparative purposes, and it is only when we wish to compare historically dated material with C-14 dated material that we have difficulties. Intuitively, bearing in mind that many of the ancient historical chronologies are based upon Egyptian records, one might feel that use of the derived Egyptian historical curve would be more meaningful as a correction method. In a sense this is of course totally so, because even if the Egyptian chronology is wrong, and thus derived Aegean or European corrections as well, it matters little as long as our *comparative* C-14 dates are converted on the currently accepted basic historical framework; that is to say, for example, that an Egyptian C-14 date of, say, 1640 BC and a European EBA date of 1640 BC both mean the same thing, namely that both samples have absolute dates of c. 1880 BC. Or one can convert both dates, using the tree-ring curve, to c. 2000 BC and still derive the one important conclusion, that the European EBA event of interest was taking place at the same time as the accession of Sesostris III to the Egyptian throne. Clearly what can *not* be done is to take the Egyptian date and, ignoring the tree-ring correction, put the event it represents at c. 1880 BC using the historical chronology, and then to correct our European EBA date on the tree-ring curve to c. 2000 BC and thus 'prove' that the European event predates the Egyptian one by some centuries: up to three centuries for some other periods. The point has been laboured but it is a real problem that is bound to increase with the improvements in laboratory dating accuracy continually being made. And there is really no best solution, so that both tree-ring and historical corrections are necessarily involved if one is to make rigorous use of all present information. It is highly probable that this unfortunately ambiguous situation will obtain for some years to come, perhaps until research with tree-ring dating of oak specimens has progressed further towards completion.

The best general correction procedure seems to be to use either appendix IB or figures 16 to 18 to yield tree-ring corrected C-14 dates, and appendix IC or figure 19 to provide comparative historical-curve corrected figures. For the period (approximately) 2450–1750 BC in C-14 years good agreement between both procedures will obtain, and before 2450 BC only tree-ring corrections are possible. But from 1750 BC until 1000 BC in C-14 years there are significant discrepancies between the historical corrections and those based on the tree-ring data. Both sets of results are necessarily to be taken into account and it would clearly be prudent for the present not to draw conclusions based upon the tree-ring corrections alone.

For practical conversion purposes the procedure recommended converts C-14 dates to a date range across the limits of the primary standard deviation, using the appropriate maximum and minimum figures in the conversion appendices IB and C. Thus, for example, a standard C-14 date of 1500±50 BC converts to a calendar date range of 2060–1700 BC using tree-ring data and 1820–1570 BC using Egyptian historical data. There is no significant variation in probability within either range, to the extent that any given year has a sensibly greater likelihood of representing the true calendar date (McKerrell 1971). Thus instead of conversion to a single year, as is the usual practice, this procedure has the advantage of demonstrating a primary feature of date correction: that the margin of uncertainty increases considerably from that of the basic C-14 standard deviation. For the examples quoted the spread of one hundred years across plus or minus one standard deviation becomes 360 and 250 years for tree-ring and historical corrections respectively, still for plus or minus only one standard deviation. There is always one chance in three that the true C-14 date lies outwith the plus or minus one standard deviation range and this uncertainty in absolute dating as induced by correction procedures is not perhaps generally appreciated, preference usually being for a central converted figure. But if such corrections are to be honestly expressed then the likely date range involved has to be stated.

Some General Implications

The discrepancies noted between historical and tree-ring C-14 dates have their major consequences after 1750 BC in C-14 years. For British prehistory this is a particularly interesting time, coming as it does at around the beginning of the Early Bronze Age and with a variety of possible connective elements towards the Continent and Near East. The two most useful general chronological outlines for this period, using standard C-14 dates, are those of Butler and van der Waals (1966) for the Central European area and Lanting and van der Waals (1972) for the British Beaker sequences. Both schemes are adequately supported by a variety of C-14 dates and need little discussion; the feature of interest is the corrected date ranges indicated. Thus conversion of the C-14 dates involved for the Butler and van der Waals outline is detailed in table 35 (and figure 20 illustrates). Table 36 lists the British Beaker steps after Lanting and van der Waals, and figure 21 illustrates the corresponding corrected date ranges.

The duration of the classical plus developed Unetice period—put by Butler and van der Waals at 200 years (periods *c* and *d*)—is reduced by the tree-ring conversion to 150 years. Bearing in mind that the converted date ranges make an allowance for the accuracy of the correction curve, and that use of the mean values instead would reduce this even further to about 110 years (appendix IB for mid-point corrections of 1600 and 1800 BC C-14 years), there is clearly a distinct problem involved. To attempt to compress this phase to a single century or even 150 years just does not seem realistic. If the classical Unetice phase was under way by c. 2200 BC in absolute years, as both tree-ring and historical corrections suggest, then it may be more reasonable to consider as likely the four centuries duration suggested by the historical corrections rather than the 100–150 years from tree-ring corrections. Period *e*, equivalent to Reinecke A2 and thus directly related to the British Wessex culture, would end by around 1900 BC in absolute years following tree-ring conversion, compared to nearer 1600 BC with historical curve correction. If we accept that the C-14 dating of Egyptian, Helladic and Minoan samples in this time span does reflect a genuine problem with regard to conversion procedures, then clearly the wholesale abandonment of the Mycenaean and other links to Britain and the Continent—as suggested by tree-ring conversions—would appear over-hasty.

The corrected Beaker dates suggest a start to the Beaker period in Britain around 2800 BC, and both tree-ring and historical conversions are essentially the same. Such a date may however be a little misleading. The eight dates from Ballynagilly average out at 1965 BC, and although there is one date of 2100 BC this seems to be due only to an entirely normal statistical spread around the mean. Similarly, the very earliest AOC sherds at Durrington Walls have a mean of 1951 BC based on four dates. These means may be a more reliable indication of the date of earliest Beakers in Britain than the maxima indicated either by Lanting and van der Waals step 1 or Ballynagilly. Step 1 for Britain could thus be 2000–1900 BC, and this in turn suggests that in absolute years, on either tree-ring or historical conversion, the Beaker period in Britain starts not much before 2500 BC. This is a little less than the earliest available Continental dates; see table 37. The time of the end of the period depends very much upon the conversion method used: the tree-ring data suggests an end to step 7 at around 1900 BC, whereas the historical conversion puts it nearer 1600 BC. Essentially this suggests either potential overlap with the Mycenaean connections of the Wessex culture, or complete incompatibility.

There are other implications of the date conversion method used. The most important is the time span over which each Beaker step lasts, and table 38 lists the relevant data. Both conversion methods increase the time spans by up to three times, but the comparison of steps 4 and 7 with steps 5 and 6 is interesting. The C-14 based chronology suggests the latter are longer than the former, but this is completely reversed by the tree-ring conversion. The historical conversion, however, maintains essentially the pattern observed with the unconverted C-14 date ranges.

There is thus, even on this brief appraisal of Beaker chronology, an indication of easier compatibility with conventional dating and historical links through use of the Egyptian based corrections rather than simply accepting the tree-ring conversions at face value.

One interesting consequence of the Central European dating is the relative situation afforded through consideration of the striking spiral and spiral-pulley designs on artifacts of known historical date and related cultures dated by C-14. Figures 22 and 23 illustrate the typical comparisons after Gimbutas (1965), and table 39 lists some of the historically dated material involved which includes examples from Mycenae, Kakovatōs, Prosymna and Tel Atchana. The Mycenae contexts are two early and spectacular shaft graves from Circle A, which could both date from as early as 1600 BC. The Tel Atchana examples relate to levels II, III, IV and VI; the latter covers the period 1750–1600 BC and is securely dated at both extremes. In all, the historically dated material falls between 1750 and 1270 BC, a span of almost five centuries. The various earliest Central European cultures involved, Veterov, Fuzesabony, Mad'arovce, Classical Otomani, etc., all relate to the end of the early bronze age and overlap into the Tumulus period. They have no connection with the Classical Unetice period and are Late Unetice or A2 in the Reinecke system. The appropriate C-14 date range is (approximately) 1600–1500 BC following the Butler and van der Waals scheme and a number of available dates confirm this. For example, at each end of the range; Bln-5047, 1370±40 BC, from an early or proto-Lusatian context (equivalent to Early Tumulus), and K-1838, 1500±100 BC from the Trzciniec culture. Direct tree-ring correction suggests a calendar date range of about 2100 to 1900 BC for this general A2 period, contrasting with 1900 to 1600 BC using the historical corrections.

Thus the tree-ring corrections put the Central European examples starting considerably before the Mycenaean or Near Eastern examples, in contrast to the conventional picture of recent years. But if the problems with correcting Eygptian and other historical dates at this point in time are meaningful, it is not necessarily appropriate to conclude that this revised picture is certainly correct. The dating overlap, if the historical corrections are used, would not clearly allow of the earliest Central European examples being significantly older than the Alalakh material, for example. The general use of such decoration from the eighteenth century BC onward (in calendar years) is thus quite appropriate for both the historically dated and the C-14 dated phases.

Two Continental phases which provide firm evidence for chronological overlap with the British Wessex culture are the Dutch Hilversum / Drackenstein culture and the Breton First Series Tumulus culture. Piggott (1973) has recently summarised the Dutch evidence; Branigan (1970) and Briard (1970) that for the striking Breton connection. Conventional C-14 dates for these phases and the Wessex culture are listed in table 40. There is little difficulty in interpretation of the Wessex and Hilversum dates and they cover the range 1470 to 1120 BC in C-14 years. But for the Breton dates there are some clear incompatibilities: one of the four Kernonen results and that for St Fiacre are respectively 1960 and 1950 BC. By any means of correction these figures suggest absolute calendrical dates in the mid-third millennium and are comparable only with the earliest of Beaker dates from Britain (for example the mean of the eight results from Ballynagilly at 1951 BC and the mean of four from Durrington Walls at 1965 BC). Moreover, the Kernonen result was (fortunately) just one of four from the same site and the other three average out at 1310 BC. Clearly there is something awry with these very old dates, and the explanation seems to lie in the fact that both samples came from wooden coffins probably carved from trees much older than the event we take them to represent. Possibly this too is the explanation for the rather old date from Lescongar since the sample here came from the massive construction timbers of the barrow. In all then there are fourteen C-14 dates to consider, and thirteen of these fall in the range 1480 to 1120 BC. It is thus, on the face of it, very difficult to see any reason for suggesting that the Wessex culture is older than 1500 BC in conventional C-14 years, and this view is well supported by an examination of the relative dating of the Wessex culture with respect to the British Beaker period. The Lanting and van der Waals chronological outline, detailed above for British Beakers, is probably the best dating scheme that we have. The dates for Ballynagilly and Durrington Walls would suggest, as mentioned above, that step 1 beakers might be no older than 2000 BC in conventional C-14 years and the complete Beaker period, steps 1 to 7, could thus cover the range 2000 to 1500 BC. In terms of the generally accepted beginning to the Wessex culture as being towards the end of the Beaker period, we could thus regard the coincidence of the two independent derivations of 1500 BC, in conventional C-14 years, as quite satisfactory. There is certainly nothing in the British Beaker dating evidence to suggest that this figure is too young for the beginnings of the Wessex culture.

Direct use of the tree-ring corrections suggest a calendrical date range of 1950 to 1350 BC, which may be compared with the historical curve range of about 1700 to 1300 BC. Since it is clear that historically dated samples do seem to diverge seriously from tree-ring results, it is appropriate to consider only this latter span for comparison with the well-known Mycenaean parallels. Moreover, the duration of the Wessex culture by the tree-ring correction procedure would be around 600 years: on the face of it a somewhat over-long span. The 400 years suggested by the historical correction is perhaps more acceptable. In all, the 1700–1300 BC date range appears the more likely and there seems no compelling reason to disregard it. It is apparent that compatibility with the Mycenaean period is independent of the means of correcting the C-14 date range involved, but that the use of the historical corrections yields a Wessex culture date range in good accord with the conventional view originally defined by Piggott (1938).

This brief review of some of the more directly affected phases suggests that the historical-curve corrections are quite meaningful, and in many cases more consistent with the archaeological evidence than direct tree-ring corrections alone, for the millennium after 2500 BC in calendar years. Until such time as the discrepancies between historical dates and corrected C-14 dates are resolved for this period, it would be prudent to consider the tree-ring calibration as tentative only.

Conclusions and Summary

A method of averaging tree-ring C-14 dates over successive fifty-year intervals of tree-rings provides correction curves that agree well with the recent MASCA data. Many of the short-term fluctuations of the La Jolla (Suess) curve are retained.

Comparison of historically dated samples and their tree-ring corrected C-14 ages yields good agreement only for the Old Kingdom of Ancient Egypt. For samples more recent than 2000 BC in calendar years there is systematic error after correction that would make the dates involved too old by up to three centuries.

For British and Continental Early Bronze Age cultures these conclusions essentially suggest that most of the difficulties involved in reconciling historically dated connective elements with corrected C-14 dates become less significant if use be made of corrections based upon Egyptian historical samples.

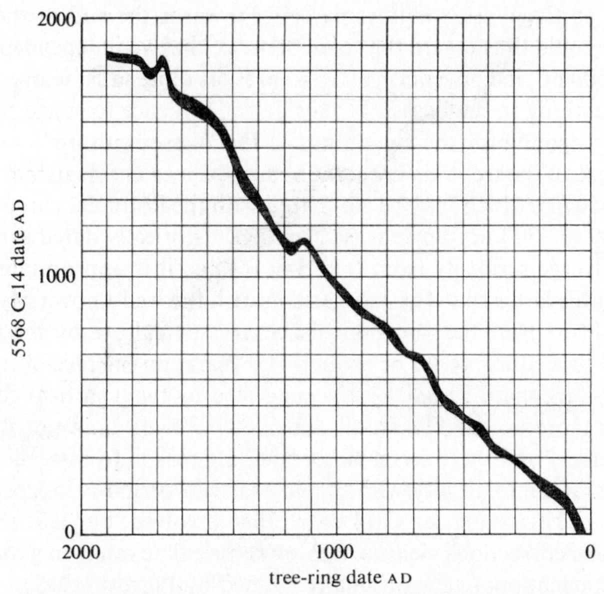

Figure 1. Tree-ring correction curve for C-14 dates averaged across fifty-year intervals of tree-ring years from 0 to 1800 AD.

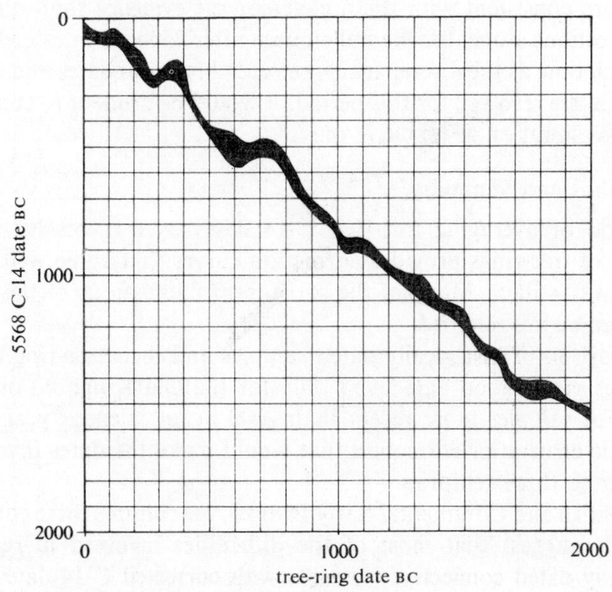

Figure 2. Tree-ring correction curve for C-14 dates averaged across fifty-year intervals of tree-ring years from 2000 to 0 BC.

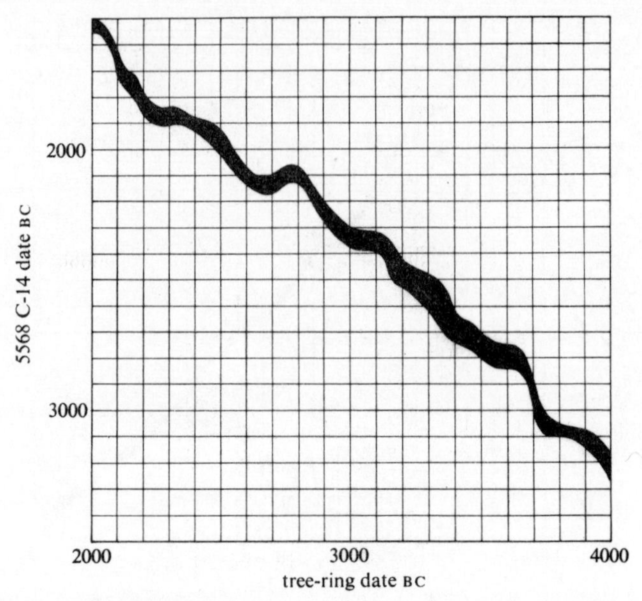

Figure 3. Tree-ring correction curve for C-14 dates averaged across fifty-year intervals of tree-ring years from 4000 to 2000 BC.

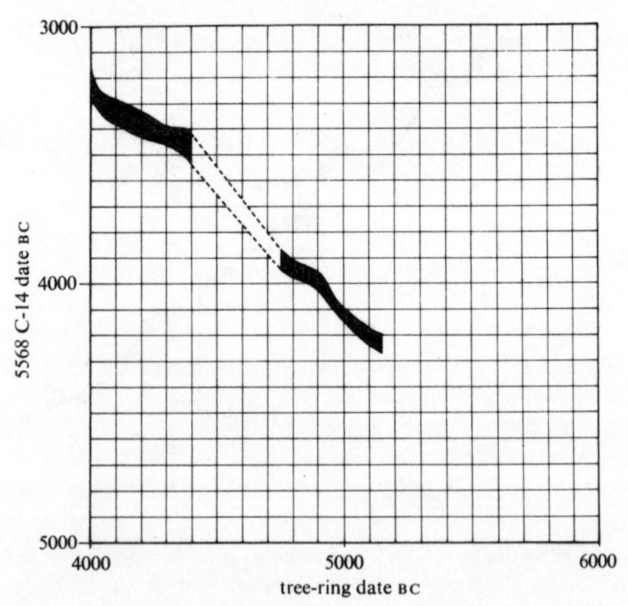

Figure 4. Tree-ring correction curve for C-14 dates averaged across fifty-year intervals of tree-ring years from 5000 to 4000 BC.

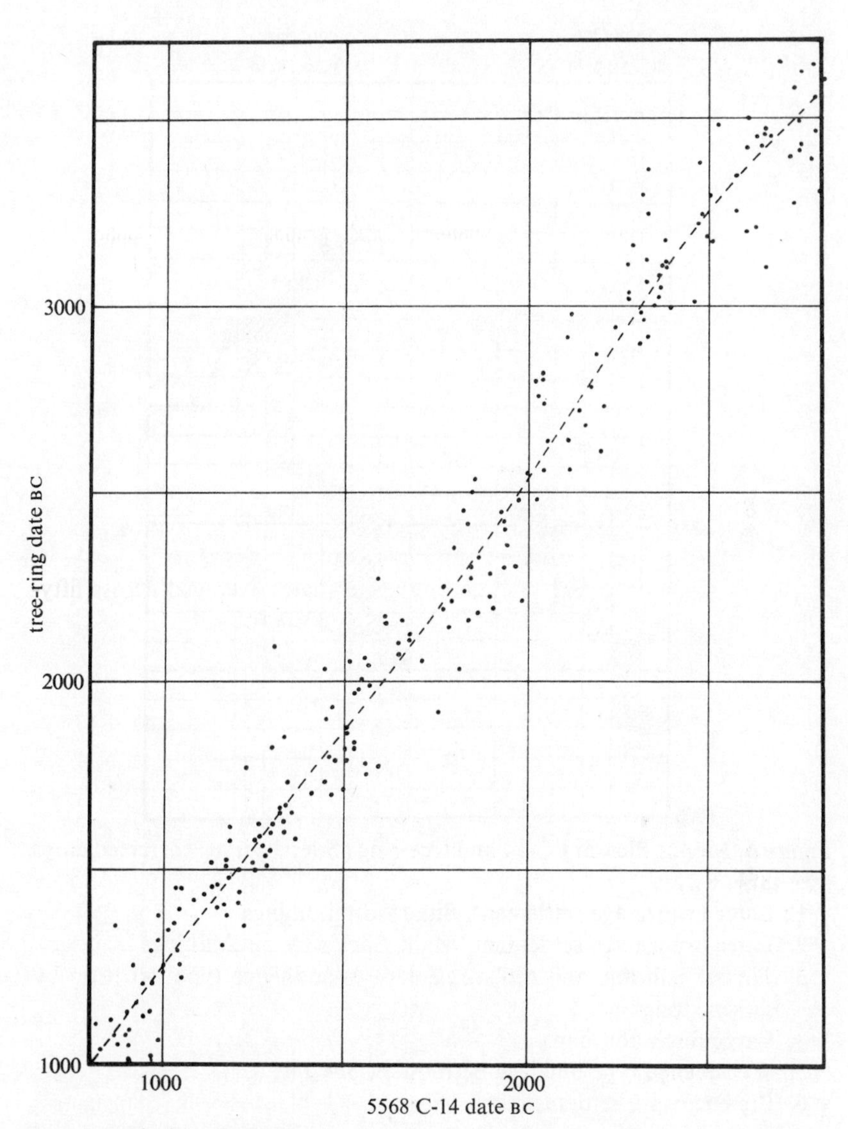

Figure 5. Sixth order regression curve for C-14 dates of tree-ring samples across 3500–1000 BC.

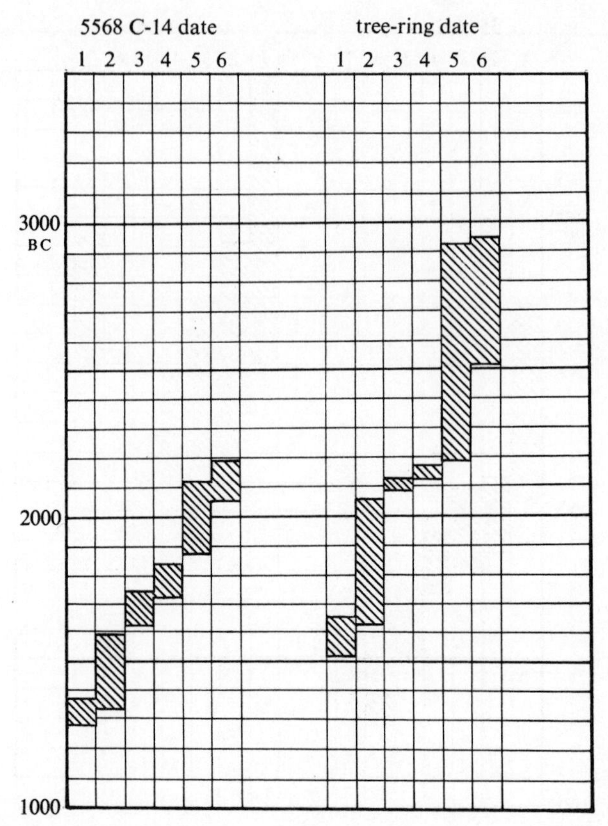

Figure 6. Mount Pleasant C-14 and tree-ring (Suess' curve) corrected dates (see table 6).

1. Later bronze age settlement, Site IV ditch fillings
2. Later bronze age settlement, Main Enclosure ditch fillings
3. Timber palisade and replacement of Woodhenge type structure by sarsen stones
4. Large ditch and bank
5. Woodhenge type building with ditch, Site IV
6. Pre-enclosure settlement.

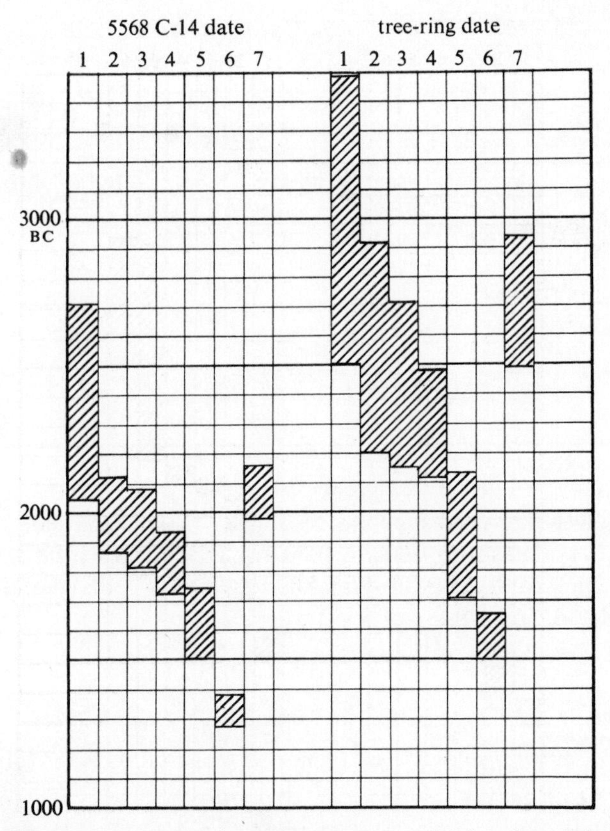

Figure 7. Henge monuments ceramic sequences, C-14 and tree-ring (Suess' curve) corrected dates (see tables 7 to 13).

1. Mid-Neolithic plain bowls
2. Grooved ware
3. Earliest beakers. Very low proportion in grooved ware material
4. Beakers and grooved ware
5. Late beakers
6. Food vessels and collared urns
7. Continental AOC beakers.

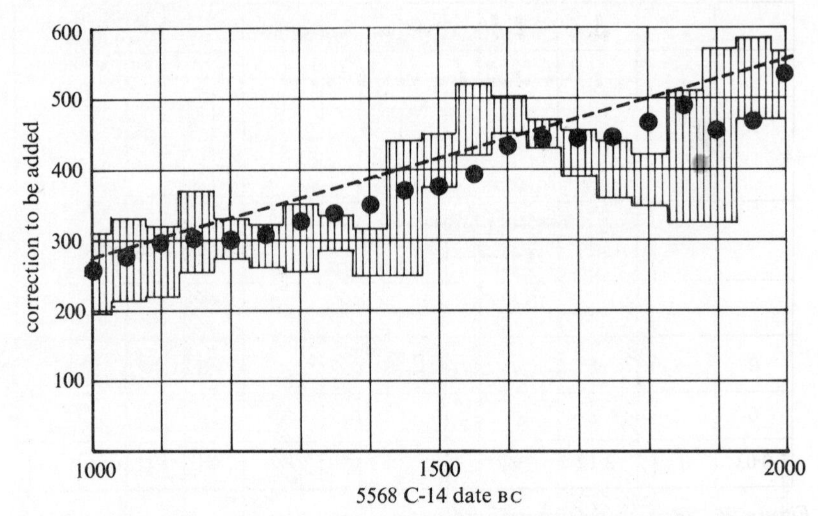

Figure 8. Tree-ring corrections to be added (making the date older) to standard C-14 dates from 2000 to 1000 BC. The hatched band is the fifty-year averaging data, the dashed line the results of Wendland and Donley, and the closed circles those of Switsur.

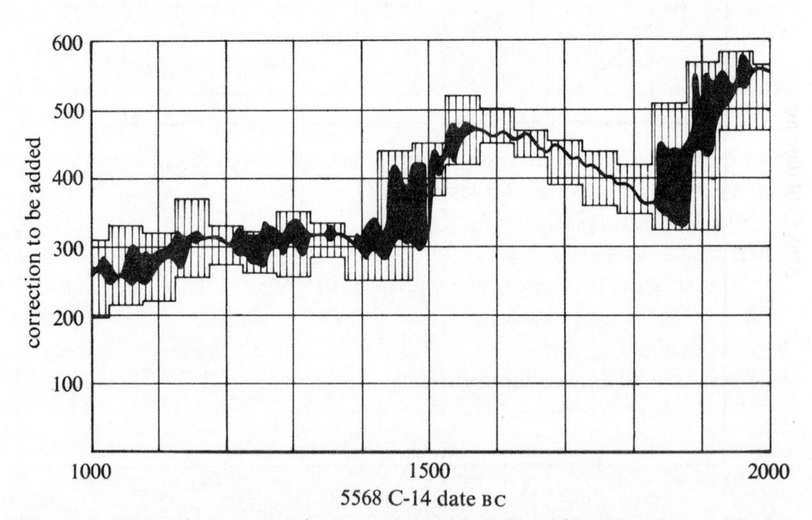

Figure 9. Tree-ring corrections to be added (making the date older) to standard C-14 dates from 2000 to 1000 BC. The hatched band is the fifty-year averaging data, the solid band the recent MASCA results.

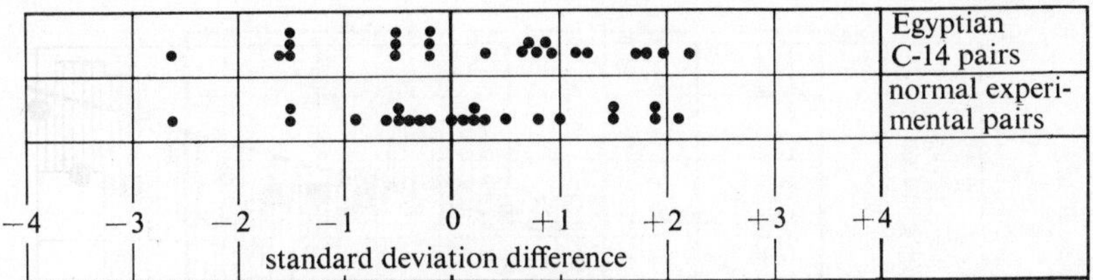

								Egyptian C-14 pairs
								normal experimental pairs
−4	−3	−2	−1	0	+1	+2	+3	+4
			standard deviation difference					
			frequency breakdown					
			11 ←	→ 12				Egyptian C-14 pairs
			10 ←	→ 13				normal experimental pairs
0	1	4	6	6	5	1	0	Egyptian C-14 pairs
0	1	2	7	7	5	1	0	normal experimental pairs
0·03	0·49	3·12	7·85	7·85	3·12	0·49	0·03	Theory

Figure 10. British Museum and UCLA interlaboratory comparison for Egyptian Series C-14 dates.

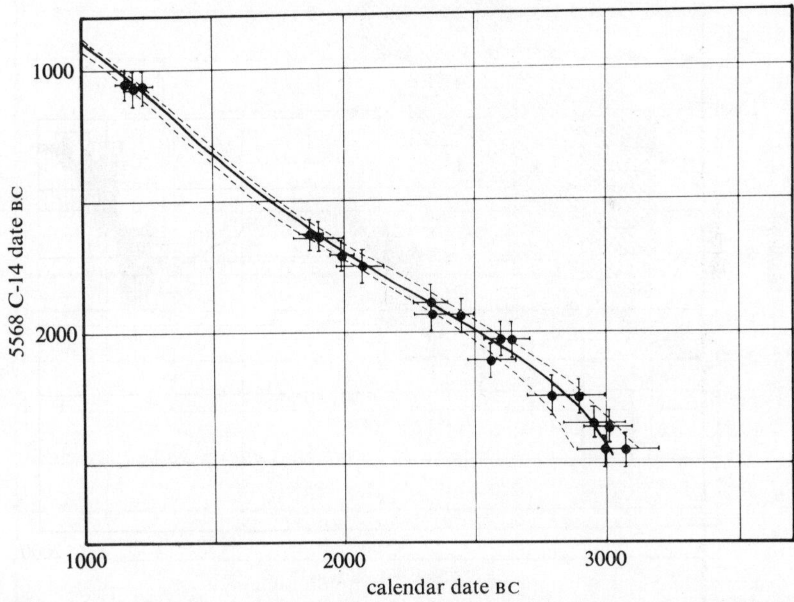

Figure 11. Fourth order regression curve for Egyptian Series C-14 dates. Each point plotted is the mean date corresponding to particular historical dates. Historical and dating accuracy are plotted for each mean.

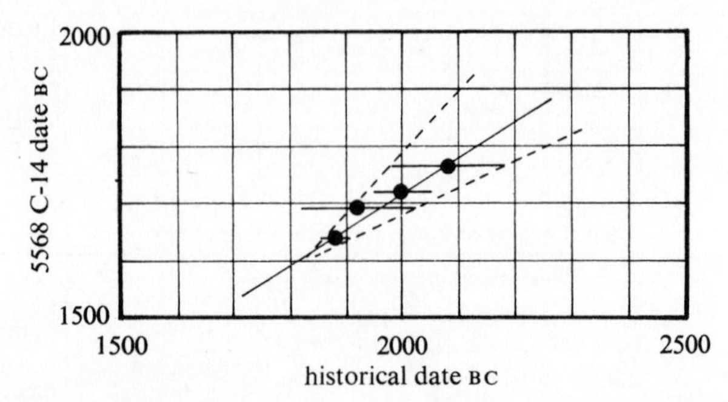

Figure 12. Egyptian Series C-14 date averages around a calendar date of 2000 BC (see table 29).

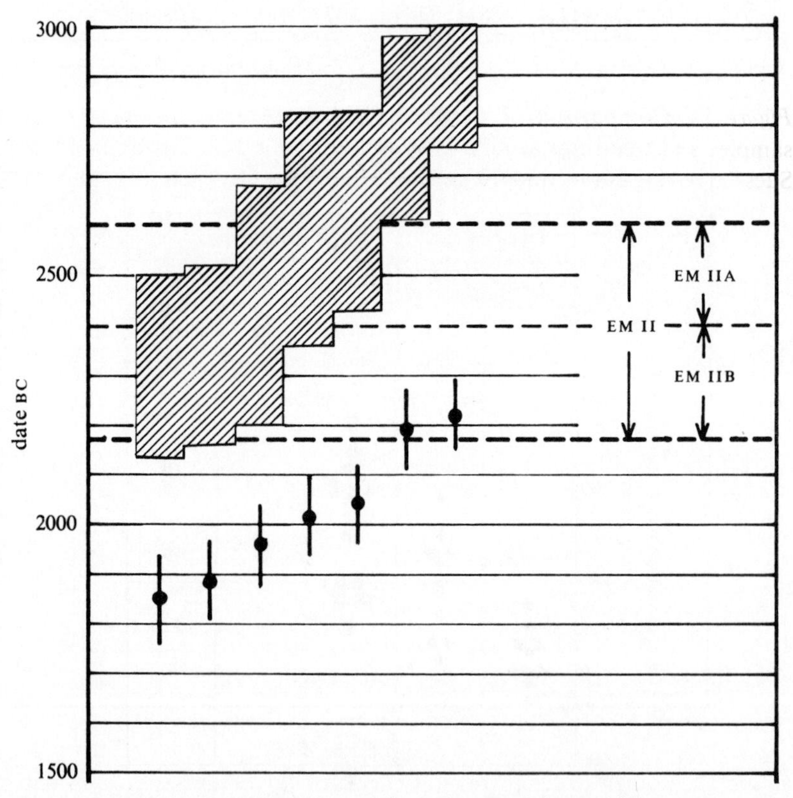

Figure 13. Standard C-14 and tree-ring corrected dates (fifty-year averaging) for Myrtos.

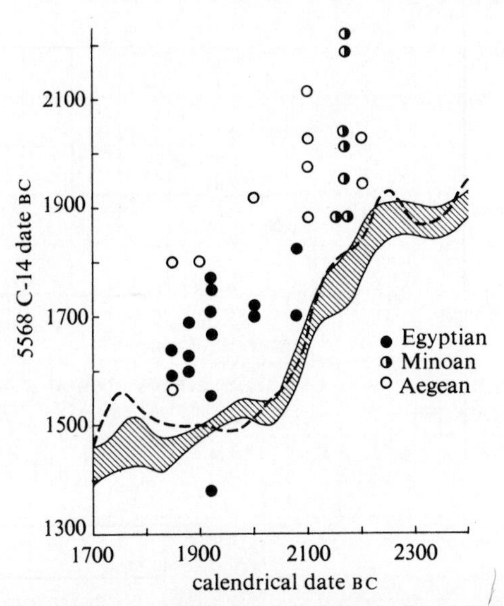

Figure 14. Comparison of standard C-14 dates from historically dated samples and tree-rings around 2000 BC (see table 34). The dashed line is Suess' tree-ring curve, and the hatched band is the fifty-year averaging data.

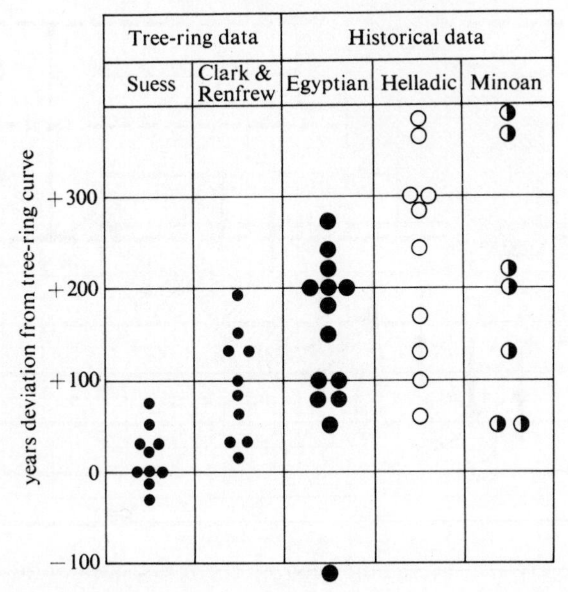

Figure 15. Comparison of the deviations from Suess' tree-ring curve of tree-ring and historical samples.

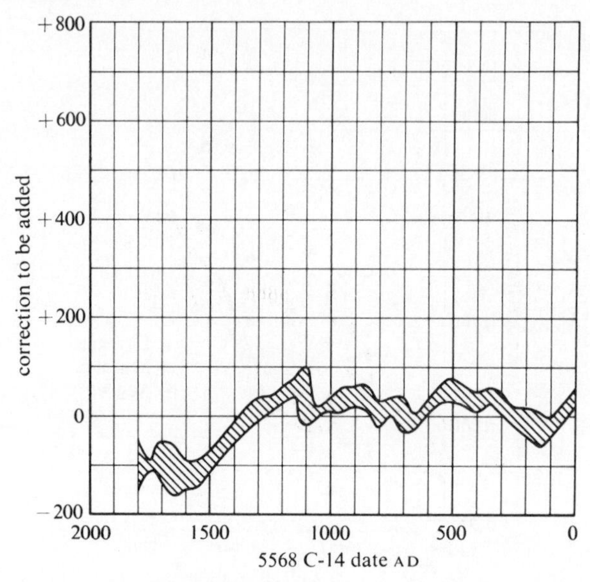

Figure 16. Fifty-year averaging tree-ring correction factors for C-14 dates from 0 to 1800 AD. The correction is added to the standard 5568 C-14 date (making the date more recent for a + correction).

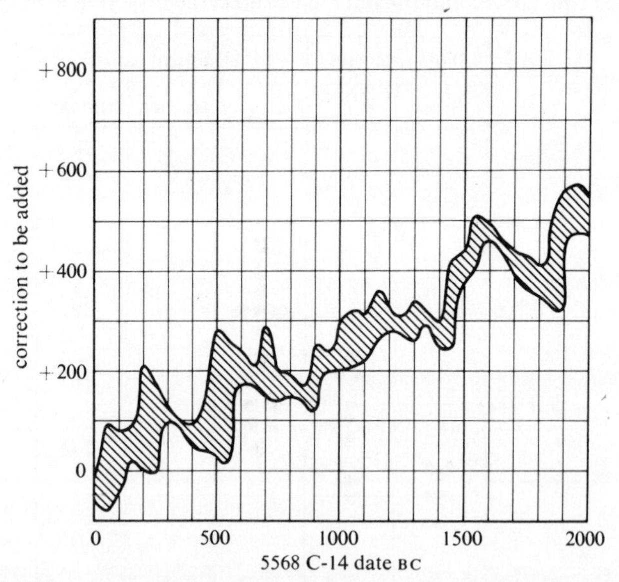

Figure 17. Fifty-year averaging tree-ring correction factors for C-14 dates from 2000 to 0 BC. The correction is added to the standard 5568 C-14 date (making the date older for a + correction).

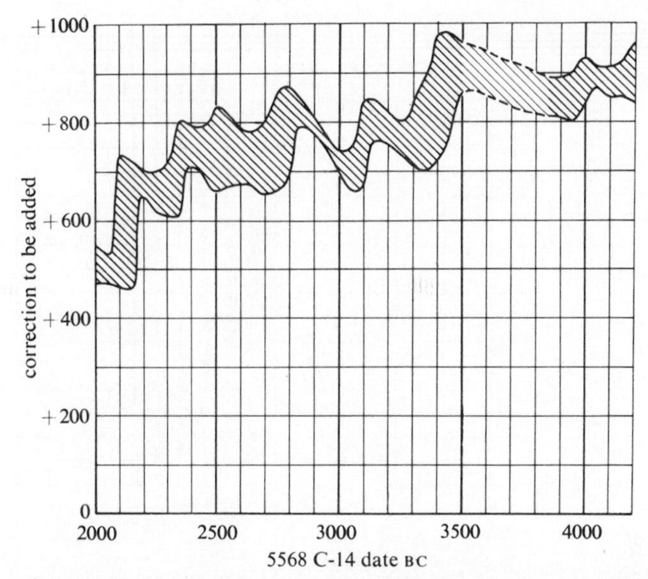

Figure 18. Fifty-year averaging tree-ring correction factors for C-14 dates from 4200 to 2000 BC. The correction is added to the standard 5568 C-14 date (making the date older for a + correction).

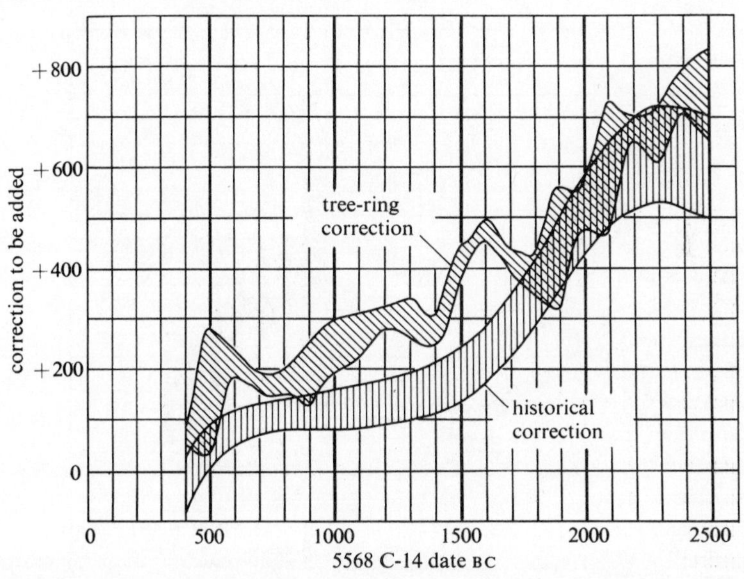

Figure 19. Fifty-year averaging tree-ring and Egyptian Historical Curve correction factors for C-14 dates from 2500 to 4000 BC. The correction is added to the standard 5568 C-14 date.

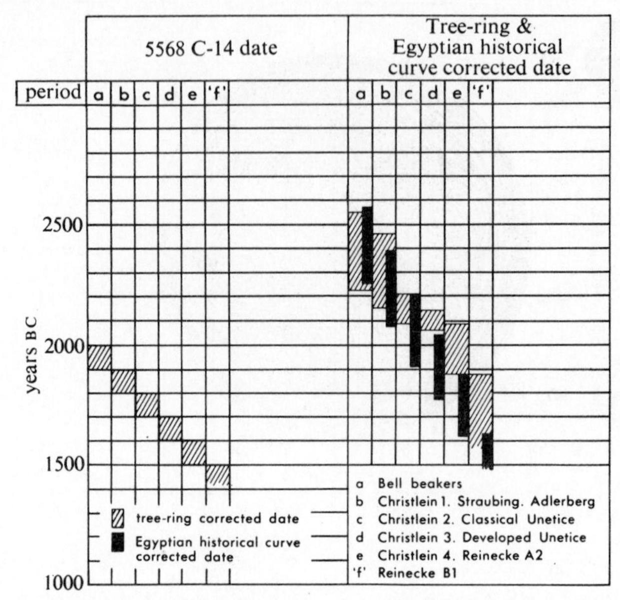

Figure 20. Standard C-14 and alternative corrected dates for central European prehistoric periods, after Butler and van der Waals.

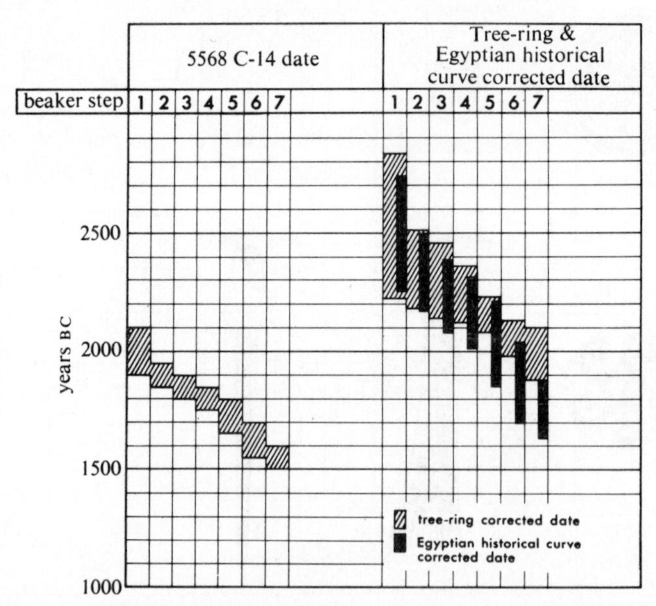

Figure 21. Standard C-14 and alternative corrected dates for British beaker steps, after Lanting and van der Waals.

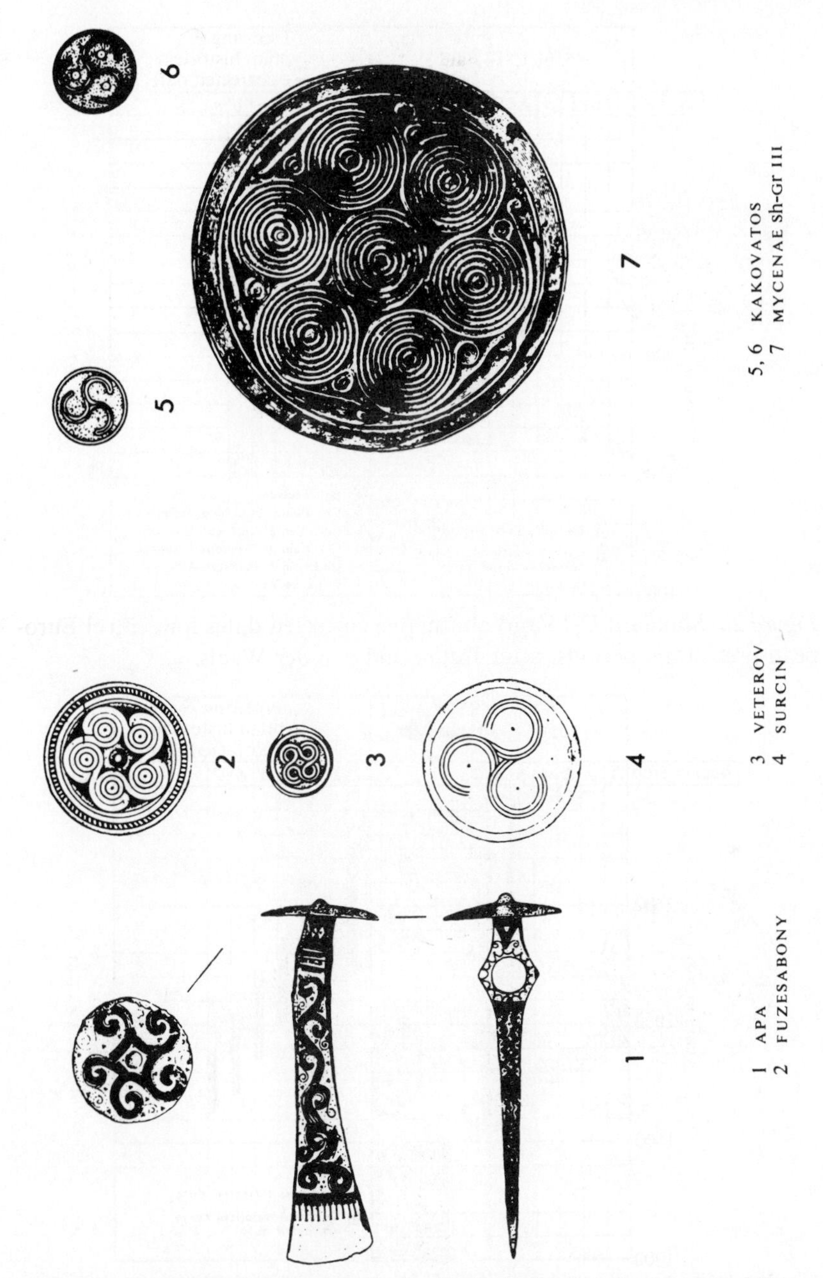

Figure 22. Spiral decoration motifs for historically dated and central European Early Bronze Age locations, after Gimbutas.

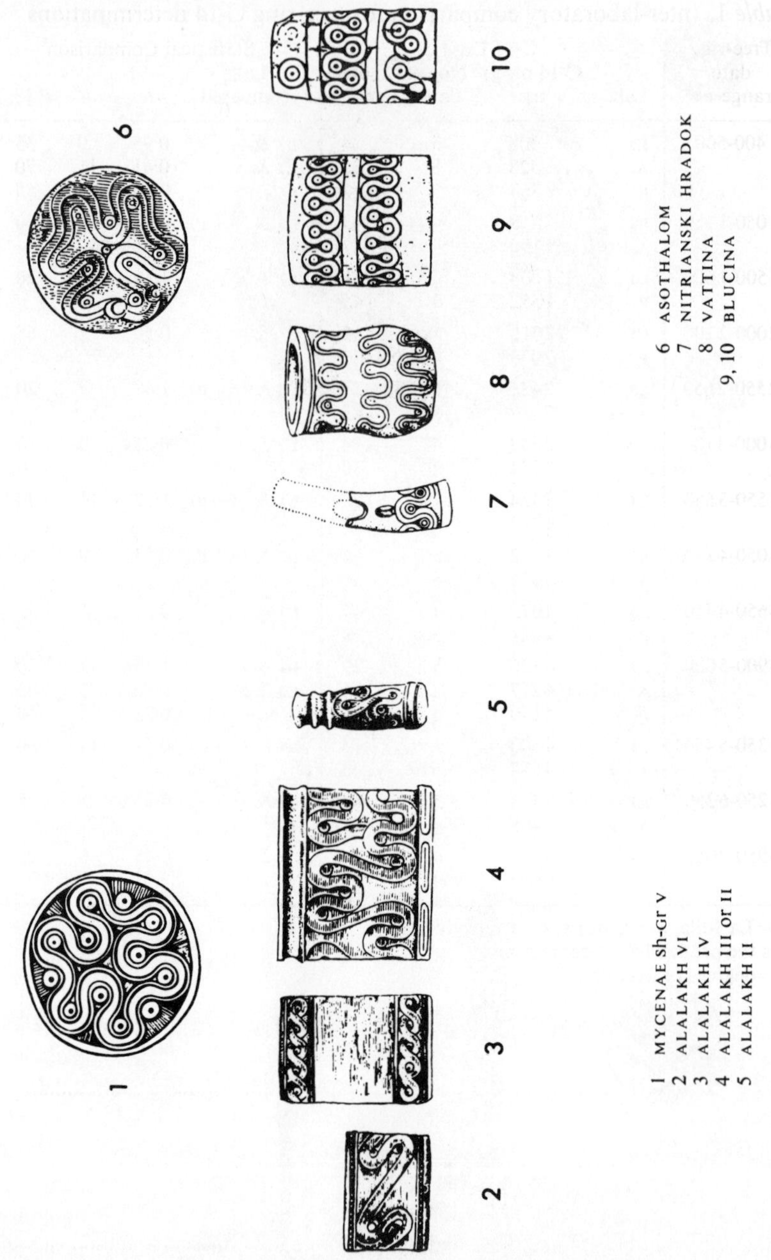

Figure 23. Spiral-pulley decoration motifs for historically dated and central European Early Bronze Age locations, after Gimbutas.

Table 1. Inter-laboratory comparison for tree-ring C-14 determinations.

Tree-ring date range BP	Lab.	C-14 mean BP	No. of dates	Std. error	Labs compared	t	df	P (%)
	C-14 Data				Statistical Comparison			
400-500	LJ	308	5	34	LJ & P	0·95	9	35
	A	328	8	32	LJ & A	0·41	11	70
	P	363	6	44	A & P	0·66	12	55
1050-1150	LJ	1118	9	15	LJ & (A+P)	0·42	11	70
	A+P	1130	4	27				
1500-1600	LJ	1608	6	15	LJ & P	1·39	8	20
	P	1552	4	45				
2000-2100	LJ	2016	5	16	LJ & P	0·50	7	65
	P	2033	4	33				
2550-2650	LJ	2425	6	15	LJ & (A+P)	1·48	9	20
	A+P	2487	5	43				
3000-3100	LJ	2843	4	24	LJ & P	0·25	9	80
	P	2854	7	30				
3550-3650	LJ	3324	7	28	LJ & (A+P)	1·12	11	30
	A+P	3272	6	38				
4050-4150	LJ	3712	5	44	LJ & (A+P)	0·53	9	60
	A+P	3674	6	54				
4650-4750	LJ	4075	4	44	LJ & P	0·88	5	45
	P	4021	3	38				
4900-5000	LJ	4322	5	29	LJ & P	1·13	7	30
	A	4277	2	36	LJ & A	0·86	5	45
	P	4250	4	62	A & P	0·28	4	80
5350-5450	LJ	4640	9	16	LJ & (A+P)	0·57	13	60
	A+P	4672	6	66				
6250-6350	LJ	5373	3	69	LJ & (A+P)	0·47	5	65
	A+P	5408	4	41				
6950-7050	LJ	6067	4	80	LJ & P	0·45	6	65
	P	6109	4	47				

LJ=La Jolla, A=Arizona, P=Pennsylvania.
P is the probability of compatibility,
df=degrees of freedom.

Table 2. Typical polynomial regression results for tree-ring C-14 dates.

5568 C-14 date	Derived Dendrochronological Date 2nd order	6th order	8th order	10th order	50/100-year averaging
1000 AD	1000 AD	1003 AD	1009 AD	993 AD	1010 AD
0	112 BC	22 BC	19 BC	4 AD	40 AD
1000 BC	1199 BC	1235 BC	1248 BC	1266 BC	1230 BC
2000 BC	2464 BC	2557 BC	2531 BC	2508 BC	2520 BC
3000 BC	3743 BC	3813 BC	3829 BC	3842 BC	3720 BC
4000 BC	5109 BC	4902 BC	4848 BC	4911 BC	4910 BC

Table 3. Date differences according to polynomial regression calculation procedure used.

5568 C-14 date	Difference (years) = [dendro. date = f(C-14 date)] − [C-14 date = f(dendro. date)] 8th order	10th order
1000 AD	+51	+46
0	−44	−52
1000 BC	+23	+30
2000 BC	−35	−49
3000 BC	+49	+45
4000 BC	−62	+2

Table 4. Minor corrected date differences between polynomial regression procedures and fifty-year averaging.

5568 C-14 date	Difference (years) = (polynomial regression − averages) 2nd order	6th order	10th order
1000 AD	−20	−17	−27
1000 BC	−152	−5	−36
2000 BC	+56	−37	+12
4000 BC	+199	+8	−1

Table 5. Major corrected date differences between polynomial regression procedures and fifty-year averaging.

5568 C-14 date	Difference (years) = (polynomial regression − averages) 2nd order	6th order	10th order
400 BC	−312	−246	−281
2100 BC	−8	−108	−74
3000 BC	−23	−93	−122

Table 6. C-14 dates for the Mount Pleasant sequences: see figure 6. (For tables 6–13 inclusive, the upper line of each entry refers to the standard 5568 C-14 date, the lower to the corresponding tree-ring corrected dates using Suess' curve. All years BC.)

Lab. Ref.	Lab. Date	−1 SD	mid.	+1 SD	Figure 6 key
BM-669	1324±51	1273	1324	1375	1
		1520	1620	1660	
BM-664	1460±131	1329	1460	1591	2
		1630	1690	2060	
BM-662	1687±63	1624	1687	1750	3
		2080	2110	2130	
BM-665	1695±43	1652	1695	1738	
		2090	2110	2130	
BM-668	1680±60	1620	1680	1740	
		2080	2100	2130	
BM-645	1784±41	1743	1784	1825	4
		2130	2140	2170	
BM-646	1778±59	1719	1778	1837	
		2120	2140	2170	
BM-663	1961±89	1872	1961	2050	5
		2180-2350	2480	2510	
BM-666	1991±72	1919	1991	2063	
		2230-2480	2490	2510	
BM-667	2038±84	1954	2038	2122	
		2380-2480	2510	2530-2930	
BM-644	2122±73	2049	2122	2195	6
		2510	2930	2940	

BM-669. Later Bronze Age settlement, Site IV ditch fillings.
BM-664. Later Bronze Age settlement, Main Enclosure ditch fillings.
BM-662, BM-665, BM-668. Timber palisade and replacement of Woodhenge type structure by sarsen stones.
BM-645, BM-646. Large ditch and bank.
BM-663, BM-666, BM-667. Woodhenge type building with ditch, Site IV.
BM-644. Pre-enclosure settlement.

Table 7. C-14 dates for Durrington Walls, Marden and Mount Pleasant Mid-Neolithic plain bowls. Plotted under column 1 in figure 7.

Lab. Ref.	Lab. Date	−1 SD	mid.	+1 SD
NPL-191	2450±125	2325	2450	2575
(D.W.)		2950	3200-3370	3380
BM-560	2654±59	2595	2654	2713
(Mar.)		3390	3400	3400-3490
BM-644	2122±73	2195	2122	2049
(M.Pl.)		2930	2530-2900	2500
NPL-192	2320±95	2225	2320	2415
(D.W.)		2930	2960	2990-3360
GrN-901	2634±80	2554	2634	2714
(D.W.)		3380	3390	3400-3490
GrN-901A	2625±50	2575	2625	2675
(D.W.)		3380	3390	3400-3470

Table 8. C-14 dates for Durrington Walls, Marden and Mount Pleasant grooved ware. Plotted under column 2 in figure 7.

Lab. Ref.	Lab. Date	−1 SD	mid.	+1 SD
BM-398	1977±90	1887	1977	2067
(D.W.)		2200-2360	2490	2510
BM-399	2015±90	1925	2015	2105
(D.W.)		2250-2480	2500	2530-2740
BM-400	2050±90	1960	2050	2140
(D.W.)		2400-2480	2510	2540-2926
BM-557	1988±48	1940	1988	2036
(Mar.)		2380-2480	2490	2500
BM-663	1961±89	1872	1961	2050
(M.Pl.)		2180-2350	2400-2480	2510
BM-666	1991±72	1919	1991	2063
(M.Pl.)		2230-2470	2490	2510
BM-667	2038±84	1954	2038	2122
(M.Pl.)		2390-2480	2510	2530-2810

Table 9. C-14 dates for Durrington Walls earliest beakers. Very low proportion in grooved ware material. Plotted under column 3 in figure 7.

Lab. Ref.	Lab. Date	−1 SD	mid.	+1 SD
NPL-240	1955±110	1845	1955	2065
		2170	2400-2480	2510
BM-395	1950±90	1860	1950	2040
		2170-2320	2390-2480	2510
BM-396	2000±90	1910	2000	2090
		2210-2470	2490	2520-2720
BM-397	1900±90	1810	1900	1990
		2160	2210-2370	2490

Table 10. C-14 dates for Woodhenge and Mount Pleasant beakers and grooved ware. Plotted under column 4 in figure 7.

Lab. Ref.	Lab. Date	−1 SD	mid.	+1 SD
BM-677	1867±74	1793	1867	1941
(Woo.)		2140	2180-2350	2380-2480
BM-678	1805±54	1751	1805	1859
(Woo.)		2130	2160	2170-2340
BM-645	1784±41	1743	1784	1825
(M.Pl.)		2130	2140	2170
BM-646	1778±59	1719	1778	1837
(M.Pl.)		2120	2140	2170

Table 11. C-14 dates for Durrington Walls late beakers. Plotted under column 5 in figure 7.

Lab. Ref.	Lab. Date	−1 SD	mid.	+1 SD
BM-285	1510±120	1490	1610	1730
		1710-1940	2080	2120
BM-286	1680±60	1620	1680	1740
		2080	2100	2130

Table 12. C-14 date for Mount Pleasant food vessels and collared vessels. Plotted under column 6 in figure 7.

Lab Ref.	Lab. Date	−1 SD	mid.	+1 SD
BM-669	1324±51	1273	1324	1375
(M.Pl.)		1510	1630	1650

Table 13. C-14 dates for Continental all-over-corded beakers. Plotted under column 7 in figure 7.

Lab. Ref.	Lab. Date	−1 SD	mid.	+1 SD
GrN-851	2190±70	2120	2190	2260
		2530-2810	2590-2930	2940
GrN-1976	2015±50	1965	2015	2065
		2480	2500	2520
Antiquity 34, 1960, 17	2145±110	2035	2145	2255
		2500	2550-2930	2930

Table 14. Egyptian C-14 dates from the BM/UCLA series excluded from analysis.

Lab. Ref.	5568 C-14 date	Reason for exclusion
UCLA-900	1691±60 BC	Potentially old wood from substantial timbers of deck of funerary boat of Sesostris III
BM-317	1483±65 BC	Collected in 1907
UCLA-739	2312±80 BC	Collected in 1914. Poor provenance
UCLA-928	2176±60 BC	Collected in 1932
BM-323	2392±70 BC	Collected in 1937
UCLA-1211	1550±60 BC	Poor provenance. Not with any certainty assigned to 11th Dynasty
UCLA-1398	1380±60 BC	The C-14 date is 8 standard errors less than the mean of five contemporaneous samples

Table 15. BM Egyptian series C-14 and historical dates (BC).

Lab. Ref.	Historical date and overall limits	5568 C-14 date and SD
BM-319	3025±100	2275±70
BM-320	3025±100	2256±80
BM-321	3025±100	2546±80
BM-322	3025±100	2399±70
BM-324	2600±80	2024±70
BM-325	2600±80	1902±80
BM-330	2335±50	1820±115
BM-331	2335±50	1820±85
BM-332	2570±50	2040±105
BM-333	1224±30	990±100
BM-334	650±40	500±70
BM-335	2000±50	1720±75
BM-336	1170±30	940±100
BM-337	1170±30	1130±75
BM-338	1200±50	1080±85
BM-340	370±30	360±80
BM-341	1920±100	1550±70
BM-342	1920±100	1710±70
BM-343	1920±100	1770±85
BM-344	600±70	660±70
BM-345	600±70	630±100
BM-346	2350±50	1910±80
BM-347	2080±100	1700±80
BM-381	600±70	593±70
BM-401	2450±50	1942±64
BM-507	2650±100	2097±60
BM-508	2650±100	2156±60
BM-509	510±5	293±60
BM-510	350±20	411±60
BM-228	3075±100	2351±65
BM-229	3000±100	2574±65
BM-230	2950±100	2429±65
BM-231	2900±100	2322±65
BM-232	2800±100	2283±65
BM-233	2650±80	2050±65
BM-234	2650±80	1836±65
BM-235	2600±80	2118±65
BM-238	1880±25	1633±65
BM-280	1880±25	1600±65

Table 16. UCLA Egyptian series C-14 and historical dates (BC).

Lab. Ref.	Historical date and overall limits	5568 C-14 date and SD
UCLA-1200	3075±100	2550±60
UCLA-1201	3000±100	2341±60
UCLA-1202	2950±100	2283±60
UCLA-1203	2900±100	2191±60
UCLA-1204	2800±100	2239±60
UCLA-1205	2650±80	2103±60
UCLA-1206	2650±80	2016±60
UCLA-1207	2600±80	2099±60
UCLA-1389	2570±50	2259±60
UCLA-1403	2350±50	1987±60
UCLA-1388	2335±50	2001±60
UCLA-1387	2335±50	1914±60
UCLA-1399	1920±100	1667±60
UCLA-1400	1920±100	1749±60
UCLA-1413	2080±100	1822±60
UCLA-1212	1880±25	1691±60
UCLA-1390	1224±30	1128±60
UCLA-1393	1170±30	1110±60
UCLA-1394	1170±30	1079±60
UCLA-1395	1200±50	929±60
UCLA-1391	650±40	579±60
UCLA-1401	600±70	642±60
UCLA-1397	370±30	385±60
UCLA-1208	2550±100	2060±60

Table 17. Historical date changes for Egyptian series.

Sample no.	Date BC as listed by Edwards	Date BC as used here	Reason for alteration
BM-333	1250	1224±30	Should date to *end* of reign of Rameses II (1290-1224 or 1304-1224 BC).
BM-338	1170	1200±50	Roma prominent during reign of Rameses II, thus his death put at only a little after that of Rameses II.
BM-341 BM-342 BM-343	2000	1920±100	Settgast (excavator) puts tomb to late 11th to late 12th Dynasties. 11th Dynasty 2134/1991 BC, 12th Dynasty 1991/1786 BC. Thus historical date of 2020 to 1820 seems best.
BM-346	2400	2350±50	Dates to time of Unas at the end of the 5th Dynasty. This would be 2350 BC.
BM-347	2000	2060±100	Dated to approximately the 11th Dynasty, 2134/1991 BC. Mid-point is 2063 BC.
BM-509	522-485	510±5	Cloth sample surrounding a document dates to year 11 of Darius I (521/486 BC).
BM-510	360-343	350±20	Mid-point used.
BM-234	2686-2613	2650±80	Mid-point used.

Table 18. Inter-laboratory comparison for BM/UCLA Egyptian series C-14 dates.

Standard 5568 C-14 dates BC				Date difference	Date difference
British Museum		University of California		(UCLA minus BM)	as a fraction
Ref. No.	Date	Ref. No.	Date	and SD	of the SD
228	2351±65	1200	2550±60	199±88	+2·2
229	2574±65	1201	2341±60	−233±88	−2·6
230	2429±65	1202	2283±60	−146±88	−1·6
231	2322±65	1203	2191±60	−131±88	−1·5
232	2283±65	1204	2239±60	− 44±88	−0·5
233	2050±65	1205	2103±60	53±88	+0·6
234	1836±65	1206	2016±60	180±88	+2·0
235	2118±65	1207	2099±60	−19±88	−0·2
332	2040±105	1389	2259±60	219±121	+1·8
346	1910±80	1403	1987±60	77±100	+0·8
331	1820±85	1388	2001±60	181±104	+1·7
330	1820±115	1387	1914±60	94±130	+0·7
342	1710±70	1399	1667±60	−43±92	−0·5
343	1770±85	1400	1749±60	−21±104	−0·2
347	1700±80	1413	1822±60	122±100	+1·2
238	1633±65	1212	1691±60	58±88	+0·7
333	990±100	1390	1128±60	138±117	+1·2
336	940±100	1393	1110±60	170±117	−1·5
337	1130±75	1394	1079±60	−51±96	−0·5
338	1080±85	1395	929±60	−151±104	−1·5
334	500±70	1391	579±60	79±92	+0·9
344	660±70	1401	642±60	−18±92	−0·2
340	360±80	1397	385±60	25±100	+0·3

Table 19. Inter-laboratory comparison breakdown for BM/UCLA Egyptian series C-14 dates.

	Data points falling into each SD range			Percentages		
SD range	C-14	Experimental	Theory	C-14	Experimental	Theory
0 to +1	6	7	7·85	52	65	68
0 to −1	6	7	7·85			
+1 to +2	6	5	3·12	39	26	27
−1 to −2	3	2	3·12			
+2 to +3	1	1	0·49	9	9	5
−2 to −3	1	1	0·49			
>+3	0	0	0·03	0	0	0·1
<−3	0	0	0·03			

Table 20. Statistical comparison of BM and UCLA Egyptian series C-14 dates.

Historical dates BC	BM dates C-14 mean BP	No. of dates	Std error of mean	UCLA dates C-14 mean BP	No. of dates	Std error of mean	Statistical Comparison *t*	*df*	*P* (%)
650 600	2546	4	35	2561	2	32	0·3	4	>50
1224 1200 1170	2985	4	43	3012	4	45	0·4	6	>50
1920 1880	3603	5	39	3652	3	24	1·0	6	~35
2350 2335	3800	3	30	3917	3	28	2·8	4	~5
2650 2600 2570	3978	8	34	4069	4	51	1·5	10	~8
3000 2950 2900	4378	4	53	4222	3	44	2·3	5	~8

P is the probability of differences being due to chance, *df*=degrees of freedom.

Table 21. Statistical comparison of tree-ring and Egyptian series C-14 dates from 600±70 BC.

Sample range BC	Tree-ring data C-14 mean BP	No. of dates	Std error of mean	Egyptian data C-14 mean BP	No. of dates	Std error of mean	Statistical Comparison *t*	*df*	*P* (%)
530-600	2469	5	26	2581	4	14	3·55	7	1
600-670	2466	6	36	2581	4	14	2·47	8	3
530-670	2451	12	26	2581	4	14	2·81	14	2
	LJ tree-ring data			A+P tree-ring data					
530-670	2457	7	17	2442	5	61	0·28	10	80

In tables 21-28 inclusive, *P* is the probability of compatibility.

Table 22. Statistical comparison of tree-ring and Egyptian series C-14 dates from 1170±30 BC.

Sample range BC	Tree-ring data C-14 mean BP	No. of dates	Std error of mean	Egyptian data C-14 mean BP	No. of dates	Std error of mean	Statistical Comparison *t*	*df*	*P* (%)
1130-1210	2887	6	12	3015	4	43	3·46	8	1

Table 23. Statistical comparison of tree-ring and Egyptian series C-14 dates from 1254 to 1140 BC.

Sample range BC	Tree-ring data C-14 mean BP	No. of dates	Std error of mean	Egyptian data C-14 mean BP	No. of dates	Std error of mean	Statistical Comparison t	df	P (%)
1140-1254	2915	11	16	2998	8	29	2·66	17	1·5
1140-1200	2889	5	14	2998	8	29	2·78	11	2
1200-1254	2930	8	19	2998	8	29	2·19	14	4
	LJ tree-ring data			A+P tree-ring data					
1140-1254	2926	6	15	2929	5	30	0·09	9	>90

Table 24. Statistical comparison of tree-ring and Egyptian series C-14 dates from 1880±25 BC.

Sample range BC	Tree-ring data C-14 mean BP	No. of dates	Std error of mean	Egyptian data C-14 mean BP	No. of dates	Std error of mean	Statistical Comparison t	df	P (%)
1880±25	3405	4	40	3591	3	27	3·58	5	1·5
1830-1880	3416	5	33	3591	3	27	3·64	6	1
1880-1930	3421	3	15	3591	3	27	5·52	4	0·5
1830-1930	3418	8	20	3591	3	27	4·69	9	0·1
	LJ tree-ring data			A+P tree-ring data					
1830-1930	3440	5	10	3380	3	49	1·58	6	18

Table 25. Statistical comparison of tree-ring and Egyptian series C-14 dates from 1920±100 BC.

Sample range BC	Tree-ring data C-14 mean BP	No. of dates	Std error of mean	Egyptian data C-14 mean BP	No. of dates	Std error of mean	Statistical Comparison t	df	P (%)
1820-1920	3398	8	29	3639	5	39	5·04	11	0·1
1870-1970	3434	5	14	3639	5	39	4·95	8	0·1
1920-2020	3468	6	18	3639	5	39	4·23	9	0·6
1820-2020	3427	13	22	3639	5	39	4·95	16	0·1
2000±25	3467	6	19	3639	5	39	4·19	9	0·4
	LJ tree-ring data			A+P tree-ring data					
1820-2020	3460	8	14	3374	5	45	2·20	11	6

Table 26. Statistical comparison of tree-ring and Egyptian series C-14 dates from 2400 to 2285 BC.

Tree-ring data			Egyptian data					
C-14 mean BP	No. of dates	Std error of mean	C-14 mean BP	No. of dates	Std error of mean	Statistical Comparison t	df	P (%)
3841	6	25	3859	6	32	0·44	10	65

Table 27. Statistical comparison of tree-ring and Egyptian series C-14 dates from 2740 to 2560 BC.

Tree-ring data			Egyptian data					
C-14 mean BP	No. of dates	Std error of mean	C-14 mean BP	No. of dates	Std error of mean	Statistical Comparison t	df	P (%)
4076	8	24	3993	6	46	1·74	12	12
LJ tree-ring data			P tree-ring data					
4088	5	35	4056	3	28	0·63	6	55

Table 28. Statistical comparison of tree-ring and Egyptian series C-14 dates from 3125 to 2900 BC.

Tree-ring data			Egyptian data					
C-14 mean BP	No. of dates	Std error of mean	C-14 mean BP	No. of dates	Std error of mean	Statistical Comparison t	df	P (%)
4293	22	24	4349	6	55	1·05	26	30
LJ tree-ring data			P tree-ring data					
4283	11	29	4318	7	61	0·58	16	55
LJ tree-ring data			A tree-ring data					
4283	11	29	4280	4	27	0·06	13	>90

Table 29. Egyptian series C-14 dates around 2000 BC in calendar years.

Historical date BC	Mean 5568 C-14 date BC	No. of dates
1880±25	1641	3
1920±100	1689	5
2000±50	1720	1
2080±100	1761	2

Table 30. C-14 dates from Myrtos.

Lab. Ref.	Historical date and overall limits BC	5568 C-14 date and SD (BC)
Q-950	2170±50 to 75	1855±85
Q-953	,,	2192±80
Q-951	,,	1885±80
Q-952	,,	2222±70
Q-1002	,,	2015±80
Q-1003	,,	1957±80
Q-1004	,,	2036±80

Table 31. C-14 dates from Lerna.

Lab. Ref.	Historical date and overall limits BC	5568 C-14 date and SD (BC)
P-303A	1850±50 to 75	1798±108
P-303	1850±50 to 75	1568±51
P-299	1900±50 to 75	1803±93
P-300	2000±50 to 75	1919±53
P-312	2100±50 to 75	1887±65
P-318	2100±50 to 75	2120±65
P-319	2100±50 to 75	2027±59
P-320	2100±50 to 75	1978±58
P-321	2200±50 to 75	1946±58
P-321A	2200±50 to 75	2031±64

Table 32. Statistical comparison of tree-ring and Lerna plus Myrtos C-14 dates for the historical period 2250 to 2150 BC.

Sample range BC	Tree-ring data C-14 mean BP	No. of dates	Std error of mean	Lerna+Myrtos data C-14 mean BP	No. of dates	Std error of mean	Statistical Comparison *t*	*df*	*P* (%)
2150-2250	3756	12	35	3969	9	40	4·00	19	0·1
	A tree-ring data			LJ+P tree-ring data					
2150-2250	3755	8	44	3759	4	66	0·05	10	90

P is the probability of compatibility, in this and table 33.

Table 33. Statistical comparison of tree-ring and Lerna plus Egyptian C-14 dates.

Sample range BC	Tree-ring data C-14 mean BP	No. of dates	Std error of mean	Lerna+Egyptian data C-14 mean BP	No. of dates	Std error of mean	Statistical Comparison *t*	*df*	*P* (%)
1850-1950	3436	6	11	3635	11	27	5·25	15	0·1
1950-2050	3478	9	15	3770	2	100	5·89	9	0·1
2050-2150	3620	8	37	3872	6	62	3·69	12	1·0

Table 34. Historical samples from around 2000 BC and related C-14 dates.

Lab. Ref.	Historical date and overall limits BC	5568 C-14 date and SD (BC)
Egyptian samples		
UCLA-900	1850±25	1691±60
UCLA-1212	1880±25	1691±60
UCLA-1413	2080±100	1822±60
UCLA-1400	1920±100	1749±60
UCLA-1399	1920±100	1667±60
UCLA-1398	1920±100	1380±60
BM-335	2000±50	1720±73
BM-347	2080±100	1700±80
BM-238	1880±25	1633±65
BM-341	1920±100	1550±70
BM-342	1920±100	1710±70
BM-343	1920±100	1770±85
BM-280	1880±25	1600±65
GrN-1177	2000±50	1705±40
GrN-1178	1850±25	1635±40
Aegean samples		
P-303A	1850±75	1798±108
P-303	1850±75	1568±51
P-299	1900±75	1803±93
P-300	2000±75	1919±53
P-312	2100±75	1887±65
P-318	2100±75	2120±65
P-319	2100±75	2027±59
P-320	2100±75	1978±58
P-321	2200±75	1946±58
P-321A	2200±75	2031±64
Minoan samples		
Q-950	2170±75	1855±85
Q-953	2170±75	2192±80
Q-951	2170±75	1885±80
Q-952	2170±75	2222±70
Q-1002	2170±75	2015±80
Q-1003	2170±75	1957±80
Q-1004	2170±75	2036±80

Table 35. Conversion of central European Early Bronze Age periods to calendar years.

Period	5568 C-14 range BC	Tree-ring corrected date range BC	Historical curve corrected date range BC
a. Bell Beakers	2000-1900	2550-2220	2570-2250
b. Christlein 1	1900-1800	2460-2150	2390-2080
c. Christlein 2	1800-1700	2210-2090	2210-1910
d. Christlein 3	1700-1600	2140-2060	2040-1770
e. Christlein 4	1600-1500	2090-1880	1880-1630

Table 36. Conversion of British beaker steps to calendar years.

Step	5568 C-14 range range BC	Tree-ring corrected date range BC	Historical curve corrected date range BC
1	2100-1900	2830-2220	2740-2250
2	1950-1850	2510-2180	2500-2170
3	1900-1800	2460-2140	2390-2080
4	1850-1750	2360-2120	2310-2010
5	1800-1650	2230-2080	2210-1850
6	1700-1550	2130-1980	2040-1700
7	1600-1500	2100-1880	1880-1630

Table 37. Earliest beaker dates for Britain. Mean or mid-point corrections.

Site	C-14 date or mean BC	Tree-ring corrected date range BC	Historical curve corrected date range BC
Durrington Walls	1951	2520-2420	2500-2340
Ballynagilly	1965	2520-2420	2500-2340

Table 38. Duration of British beaker steps.

Step	5568 C-14 based duration	Tree-ring corrected based duration	Historical curve based duration
1	200	610	490
2	100	330	330
3	100	320	310
4	100	240	300
5	150	150	360
6	150	150	340
7	100	220	250

Table 39. Spiral-pulley and spiral designs.

Site	Cultural Group	Site	Date Range BC
Central European spiral-pulley motif		*Historically dated spiral-pulley motif*	
Asothalom	Fuzesabony	Mycenae. Shaft grave V	1600-1525
Vattina	Vattina	Atchana II	1350-1270
Nitrianski Hradok	Mad'arovce	Atchana III or II	1370-1270
Blucina	Veterov	Atchana IV	1450-1370
		Atchana VI	1750-1600
		Prosymna (LH II-IIIA)	1500-1300
Central European spiral motif		*Historically dated spiral motif*	
Veterov	Veterov	Kakovatos	1500-1450
Fuzesabony	Fuzesabony	Mycenae. Shaft grave III	1600-1525
Surcin	Vattina		
Apa	Classical Otomani		

Table 40. Standard C-14 dates BC for the Wessex culture and directly related Continental phases.

Wessex		
BM-680	Earls Barton	1219±51
BM-681	Earls Barton	1264±64
BM-682	Hove	1239±46
BM-708	Edmonsham	1119±45
Hilversum/Drachenstein		
GrN-1028	Knegsel	1140±30
GrN-1828	Totefout	1470±45
GrN-2968	Treek	1380±70
GrN-2997	Vogelenzang	1139±70
GrN-5716	Nijnsel	1140±75
Breton First Series		
Goy-235	Cruguel	1320±200
Gif-806	Kernonen	1250±120
Gif-807	Kernonen	1200±120
Gif-1149	Kernonen	1480±120
Gif-805	Kernonen	1960±120
Gif-683	St Fiacre	1950±135
Gif-749	Lescongar	1620±115

The Implications of Calibration

A. FLEMING

This paper documents the reaction of one prehistorian to the questions raised by the calibration issue. It is not based on detailed research into the problems of calibration; if it has value this will be because of its relevance at this moment in time to some contemporary archaeological thought, and perhaps because of one or two general questions raised. I want first to address some remarks to the scholars working in the interface area between archaeology, physics and statistics, before going on to discuss the wider implications of calibration.

We are, unfortunately, on the verge of anarchy when it comes to the question of whether, when and how to calibrate C-14 determinations. At one end of the scale there are prehistorians like me who feel that little should be done until there is international agreement on the subject; at the opposite extreme, some museum curators have already changed their labels. This is an appalling situation; it reflects no credit on us, and makes the task of anyone teaching undergraduates or evening-class students a very difficult one. At the moment any prehistorian attempting to calibrate his dates faces formidable problems. Which curve is he to use? What does he have to do to take the new half-life into account? Supposing there are short-term fluctuations in the production of atmospheric C-14, superimposed on some more long-term trend, what element of uncertainty or imprecision does this introduce? Is increased imprecision a feature of the new situation, and is there a standard way in which one may allow for this? Are there periods in which the imprecision becomes greater?

Another problem is that of differential coverage. For the whole of the Upper Palaeolithic and most of the Mesolithic periods we cannot calibrate our dates because of the age limits for dendrochronology; for the Bronze and Iron Ages there is little to calibrate as yet, because prehistorians have assumed that the historically-derived chronology was correct and probably more accurate than C-14 dating, and hence have obtained few determinations.

In my opinion there is only one reasonable course of action for prehistorians to take; it is to honour existing agreements, and quote dates in C-14 years on the old half-life. After all, it is only for those interested in the relations between different culture areas that absolute chronology is of major importance, and even in this field the extent of the problems posed by calibration should not be exaggerated. We should try to avoid conflating three basic areas of study: C-14 dating, derived historical dating of the Wessex-Mycenae type, and the study of similarities between artifacts. As C-14 dating increases in importance and sophistication, derived historical dating will become unnecessary. To try to use the two together is dangerous, since the assumptions which underpin them could hardly be more different. As for those who continue to interest themselves in similarities and possible links between different areas, one may legitimately ask them what they are seeking, now that it can no longer be chronological information. I will have more to say on this point later.

It is clear, however, that some international agreement on calibration should be reached as soon as physicists and statisticians feel that this is possible; it is clearly not for archaeologists to try to stampede them into premature action. It may be that it will be twenty years before such an agreement is possible; if that is so, it is worth considering some sort of provisional agreement, perhaps for a restricted area like the United Kingdom or Europe. Ideally we should be given clear instructions about how to calibrate our dates by the scientific community which produces them; a lesser alternative, but still one preferable to anarchy, would be an operational agreement among prehistorians, in which pragmatic considerations, rather than the niceties of physics, would be the principle moving force. Whatever happens, it is important to have clear leadership on the *mechanics* of calibration as well as on its implications.

Renfrew has pointed out some of the chronological implications of calibration, and I agree with him that it is the methodological implications which are most interesting. But let us look briefly at the changed chronological picture, much of which, after all, depends more on the existence of C-14 dating than on calibration. Conventional C-14 dates were causing problems before calibration became an issue. For instance, the beginnings of metal-working in the Balkan area were already dated to at least 3500 bc (C-14 years) and it was very difficult to find prototypes for the elegant axe-adzes in areas further east. Now, in the widening gap which opens up near the beginning of the calibration curve, we may be dealing with a dating of nearly 5000 BC, and it is increasingly tempting to see an independent origin of copper-working in south-east Europe. Then we have to consider the origins of the great monumental stone tombs of western and northern Europe. Some of the Breton sites were already dated to 3500 bc or earlier;

the earliest ones now date back to nearly 5000 BC. Even before calibration these tombs clearly had origins which had nothing to do with the east Mediterranean. In any case, convincing morphological parallels have never been found in this area. The British long barrows, monuments implying comparable organisation and skill, also move back into the fifth millennium, and no Mediterranean parallels have been claimed for these. So the case for vigorous cultural evolution in prehistoric Europe does not *depend* on calibration, and this in itself predisposes one to accept Renfrew's suggestions made for times when calibration *is* critical.

Calibration does become critical for C-14 dates in the third millennium (in C-14 years). These affect many west European megalithic tombs, the Maltese temples, and the fortified Copper Age sites in Spain and Portugal. If these sites really date from the third millennium, east Mediterranean prototypes for them and the cultural material associated with them are implausible rather than impossible; calibration forces one to confront the issue, either by searching for fourth-millennium links or accepting a more independent character for European cultural development. Some prehistorians have questioned the notion of missionaries, colonists, metal-prospectors or travelling undertakers sweeping through the straits of Gibraltar imposing mother-goddess worship on the hapless natives. The idea has been criticised on archaeological grounds as well as in terms of general implausibility. Thus independent development, long suspected for reasons which had nothing to do with C-14, is made more or less a certainty by calibration.

In the second millennium prehistorians have traditionally seen a horizon of Mycenaean trade, or at any rate influence of some kind, and of course they have had a vested interest in doing so in order to build up any chronology at all. What has happened in the case of the Mycenaean horizon is very unfortunate. So many prehistorians believed that the occurrence of faience beads carried a close and precise dating that they neglected to submit samples from the European Early Bronze Age for C-14 dating; when Renfrew came to consider the matter he had to use the Stonehenge dates, and of course the relationship between the Stonehenge building phases and the faience horizon is not known. It was 'Stonehenge without Mycenae', rather than 'Wessex without Mycenae', and in any case the original suggestion about the possibility of a Mycenaean architect for Stonehenge was surely intended as a flight of historical imagination, a possibility rather than a probability. Since then there has been debate about the place of manufacture of the British faience beads; it is unfortunate that an argument about chronology and calibration should have become entangled with a separate one about characterisation. One can only say that if calibration is accepted, it looks as if the latter part of the Wessex culture and the last phase of Stonehenge *were* contemporary with Mycenae, and that this should not affect our judgement of where the faience was made, which is an altogether different question. But it seems that the more magnificent early phases of Stonehenge (II and IIIA) are pre-Mycenae. As for the early part of the Wessex culture, we await an adequate spread of dates. Early dating need cause no concern. We know that brilliant civil engineering at Silbury Hill occurs before the pyramids and that complicated carpentry in heavy timber at Durrington Walls precedes Mycenae very comfortably. The less convincing Mycenaean links in the British Early Bronze Age have been under attack for some years. All these things certainly have the effect of sending the Mycenaean architect back to his drawing-board, as it were.

At this point the rarity of C-14 dates makes it hard to carry on with our revised chronology, while the narrowing gap between calibrated and uncalibrated dates makes the revisions less drastic anyway. We are now in the period between the alleged Mycenaean horizon and the horizon of Greek and Mediterranean imports in the European Iron Age. However, even in this period calibration will have some effect; for instance calibration of the dates from sites like Ram's Hill (Berkshire) and Mam Tor (Derbyshire) will tend to push the origins of hill-top settlements and incipient fortifications further back, so that the British examples may well precede the European.

One would also like to know how far the apparent increased imprecision of a calibrated chronology will affect some other crucial issues. For instance, what will be the effect on the thinking of palaeobotanists, who have to decide how far the elm decline corresponds to a restricted chronological horizon, and how far it is scattered across a broad chronological band? Increased imprecision may make it more difficult to separate climatic influences from the changes wrought by man. The same problem may affect the argument about the formation of peat deposits in upland Britain. Increased imprecision will not help to decide how far peat formation is strictly a local matter, or how far it relates to phases of increasing wetness at specific times such as the onset of Atlantic or sub-Atlantic climatic conditions. Those pollen analysts who are attempting to look at monoliths in a very detailed way, in order to find out more about the nature and duration of clearance episodes, will not find increased imprecision helpful either.

When archaeologists discuss chronological sequences in detail, they often use them in arguments about causality. They are fond of asserting that, because B follows A, there may be a causal relationship as well. At any rate, the establishment of the true order of events is a vital basic step and, at the moment, if two C-14 dates are separated by two or three hundred years in the Bronze Age or the Neolithic, this is taken to mean that for practical purposes they are statistically separate. If calibration means that this margin has to double, and that dates in general have a much broader frame of reference than has been believed up to now, our simplistic argument about chronological priority and causation will become very difficult if not impossible. We may have to eliminate certain policies for obtaining determinations as simply not worthwhile; for instance, the practice of the 1960s of obtaining *one* date per long barrow. Prehistoric archaeologists would welcome guidance on these matters. It may be, that in places where the curve becomes most vague, that both our view of the accuracy of our chronological picture *and* our sampling policies will have to change drastically.

So if we accept the need to calibrate or at least to follow the implications of calibration, we find that it confirms the suspicions of those who have never sympathised with a view of prehistoric Europe steeped in outer darkness waiting for inspiration from the east. The prehistoric Europeans were evidently more inventive than was thought a decade or two ago. Now we may react to this situation in two ways. The first is to see the argument simply in terms of changing fashions in the old Diffusion versus Independent Invention controversy, to observe that diffusion is going out of fashion, and then perhaps to display amused tolerance of the antics of the young. This approach is rather too cynical. It is indeed true that since we know from history and ethnography that diffusion and local invention do occur and have occurred in many present and past situations, there is no particular virtue in an instinctive preference for one or the other as explanatory mechanisms. We cannot say 'this must be local in origin, unless proved otherwise' any more logically than we can say 'this looks new, so it must have come in from outside'. In fact, the whole controversy has outlived its usefulness. We need to question our first habit of asking 'where does this come from?' and treating the answer to that question as one of our major concerns. Why, in any case, are we so keen to identify the bearers of culture? 'Whodunnit?' is a more appropriate question for the detective than for the prehistorian, when one comes to consider the matter.

In one or two recent books it has been suggested that races differ in mean intelligence, and that these differences affect their relative capacity to develop civilisation. Now I am not competent to comment on remarks made about the 'intelligence' of different racial groups, but it does seem remarkable that even a few people should seriously believe that the stage of culture reached by a population has anything to do with the average intelligence of its members. To believe this also implies the belief that many of the world's peoples, past and present, were and are too stupid to develop their own cultures but bright enough to see the advantages of adopting new inventions once they have been introduced. So there are some groups whose inventors come from within their own societies, by and large, and other groups who have to wait for stimuli from outside. The first groups respond to local initiatives, the others only to external ones. Even supposing that this situation is true—which I do not accept—it is arguable that peoples who have the capacity to be *receptive* to new ideas, and whose members are flexible enough to see the utility of new inventions and discoveries, are behaving more intelligently from the evolutionary point of view than those who seek, as 'civilised' peoples have done, to export their own cultural traits beyond their natural environment. In any case, the members of highly developed cultures have already demonstrated their own receptivity to cultural change. I am not suggesting that there are many prehistorians who do award bonus points to those past peoples who seem more inventive than others (although Piggott has sought to distinguish between 'innovating' and 'conserving' societies); but I think that we have become too preoccupied with the 'who were the culture bearers?' type of question.

As prehistoric archaeologists we are in any case poorly equipped to discuss this question. For instance, diffusion is a very difficult concept for us to handle. If we are honest, we are trying to retrace the steps of individuals or small groups of people. Occasionally we may do this quite successfully, with craft objects, or by tracing the geological origin of the tempering material used in pottery. But most diffusionist arguments depend on general similarities between pots or arrowheads or buildings; and the most deceptive arguments of all are those which use the patterns on distribution maps as records of the movement of people, without regard to time depth, differential archaeological endeavour, and so forth. Archaeologists, it is my impression, fall into two groups when it comes to diffusion; some think in very concrete terms of smiths and traders, others like to keep the concept as abstract as possible to avoid having to think about what may actually have happened. Ethnography might be expected to be useful, but unfortunately most documented cases of diffusion relate to European contact situations. Here, in view of the gap between colonials and natives, the 'culture bearers' model may be appropriate, but it will be far less so when the differences between the two groups in contact are slight. It may be that, almost by definition, Europeans have never witnessed and recorded such cases of diffusion and never will do.

While we are discussing innovating and conserving societies, it would be interesting to know whether anyone has ever taken up Childe's suggestion about plotting the invention rates of different societies on a graph, and whether there are variations from group to group. I suspect that if this were done there would indeed be great variation; after all, when a society has reached what is sometimes termed a 'stable adjustment' it has no need for a high innovation rate, whereas one which is in the process of adapting to its environment or moving towards internal equilibrium may become more inventive. And one would expect the invention rate to increase generally through time as society and economy increase in complexity; incidentally, calibration, which has the effect of lengthening the earlier phases of farming and metallurgy, is tending to reinforce this impression.

So I believe that it is pointless to carry on the discussion on the diffusion-independent invention continuum. It is more important to try to understand the role of a new development in a society than to ask precisely where and how that development took place, questions which are probably incapable of firm resolution or further extension. One may use chamber tombs as an example. We know that they vary a great deal. Some tombs are hardly anything more than large stone boxes for a selection of the dead; others are grandiose ceremonial monuments, designed as much to impress the living as to contain corpses. I have suggested that the latter may have played a critical role in sustaining the more developed societies which must have accompanied the spread of farming in western Europe. It may be that local groups in a dispersed settlement pattern needed a focus to prevent tendencies towards fragmentation; that as the group became larger, authority needed reinforcement; or perhaps if status had a hereditary basis symbolic validation was important. These ideas are frankly speculative but they do lead us to ask further questions, some of which are at any rate answerable. We want to know in what areas these tombs coincided with the earliest farming, in what areas they related to a well-established, mature culture. We need to separate the more imposing from the more humdrum tombs, and see whether they correlate with particular types of environment. We may need to ask physical anthropologists to re-examine the whole question of how far family relationships may be inferred from bones. If tombs really were designed to impress, can their siting factors be studied in more detail?

As a result of calibration, we can no longer avoid the issue of how far we believe in evolutionary concepts in prehistory. At the moment, prehistorians in Europe have not by and large felt it necessary to go into the debate between the Boasians and anthropologists like White, Service or Steward. Yet the discussion has great relevance; prehistorians have taken a Boasian position often without being aware of the counter-arguments. In European prehistory this has meant a view of the past as a series of events, or fashion changes; in our retention of the Three Age System we have accepted the evolution of increasingly improved materials for cutting tools, but there has been little attempt to trace evolution of, as opposed to changes in, such things as subsistence techniques, ceremonial behaviour, social organisation, settlement types, and so on. We have totally failed to examine the question. We do not know whether the recorded changes in these things are disconnected events or part of coherent evolutionary patterns, and this is largely because we have become preoccupied with tracing origin areas. Now that invasions have become unfashionable, there is an increasing sense of continuity in British prehistory; let us examine this continuity further, to see whether it betokens evolutionary development. There is no space to develop this theme further. But it must be accepted that calibration does *not* mean simply a few hundred years more on our dates, nor the rapid abandonment of such spectres as megalithic missionaries and Mycenaean faience-pushers; ultimately it means a wholesale re-examination of our approach to our own past. The issues were there before calibration became a talking-point, and they will still be there when some sort of calibration agreement has been reached; what seems a technicality may have forced us to confront them for the first time.

Appendixes

Appendix I
Conversion Tables

H. MCKERRELL

Appendix I A

Conversion of standard (5568) C-14 *dates to tree-ring calendar years: Comparative data*

Comparison of the various published graphs or lists for converting C-14 dates to tree-ring calendar years is possible, at fifty-year intervals of standard C-14 dates, using these tables. The sources of the data used are detailed below.

5568 C-14 *date*. The standard C-14 date using the 5568 year (Libby) half-life. This is the system used in *Radiocarbon* and would be the quoted archaeological C-14 date.

5730 C-14 *date*. The converted C-14 date using the present best estimate of the half-life of C-14 (5730 years against 5568 as above). See Godwin (1972).

Suess (*1970*). Data derived by extrapolation from the curve published by Suess in the Nobel Symposium. See Suess (1970).

Wendland and Donley (*1971*). A third order fit by regression analysis of all the (then) available tree-ring data. It would be expected that this technique should perhaps oversmooth the curve and indeed almost a straight line is obtained. See Wendland and Donley (1971).

Switsur (*1973*). The average of corrected dates from the data presented at the New Zealand Radiocarbon conference in 1972 by the Pennsylvania and Arizona laboratories. The Pennsylvania work used an averaging technique, the Arizona work polynomial regression. See Switsur (1973). Uses all published tree-ring data.

MASCA (*1973*). The most recent assessment of the data by the Pennsylvania laboratory. Based upon an averaging technique. See Ralph, E.K. *et al.* (1973). Uses all published tree-ring data but no estimate of overall reliability.

50-year averaging (*1974*). A similar procedure to that used by MASCA (1973) but based upon the average C-14 date across successive fifty-year intervals of tree-ring span. Uses all published tree-ring data, and the range tabulated is based upon the standard error of each fifty-year mean.

Egyptian historical (*1974*). Based upon a correction curve derived from accurately dated Egyptian historical samples from 3100–400 BC (see Edwards 1970, Berger 1970). A few dubious samples were rejected and the rest fitted to a low order polynomial curve across the limits of historical accuracy. The range tabulated thus covers the likely accuracy involved.

5568 C-14 date	5730 C-14 date	Suess	Wendland & Donley	Switsur	MASCA	50-year averaging	Egyptian historical
1800 AD	1796 AD	1660-1760 AD	1728 AD	1735 AD	1670-1770 AD	1650-1760 AD	
1750	1744	1650	1690	1666	1650	1640-1660	
1700	1692	1510-1640	1651	1628	1530-1630	1570-1650	
1650	1641	1500-1630	1611	1582	1510-1600	1490-1590	
1600	1590	1480	1570	1540	1470-1500	1450-1510	
1550	1538	1450	1529	1516	1440	1410-1460	
1500	1487	1430	1487	1479	1420	1390-1430	
1450	1436	1400	1444	1419	1410	1370-1410	
1400	1384	1370	1401	1377	1380	1350-1390	
1350	1333	1330	1357	1332	1340	1320-1370	
1300	1281	1290	1313	1293	1300	1290-1340	
1250	1230	1260	1267	1257	1250	1260-1290	
1200	1178	1230	1221	1214	1220	1230-1260	
1150	1127	1210	1175	1169	1190	1190-1230	
1100	1075	1130-1210	1127	1119	1110-1140	1080-1200	
1050	1024	1070-1210	1080	1069	1060	1050-1070	
1000	972	1030	1031	1021	1020	1010-1040	
950	921	1010	982	974	980	960-1010	
900	870	970	932	926	940	920- 960	
850	818	860	882	869	890	860- 910	
800	767	840	832	820	830- 850	780- 830	
750	715	770- 830	780	780	770- 790	750- 790	
700	664	710	728	729	700- 720	670- 740	
650	612	670	676	673	650- 670	630- 660	
600	561	640	623	625	620- 640	610- 640	
550	509	610	570	598	590	580- 620	
500	458	570	516	536	550	530- 580	
450	407	530	462	484	490- 510	470- 510	
400	355	440	407	424	440	410- 450	
350	304	410	352	367	390	380- 410	
300	252	330	296	317	290- 320	300- 350	
250	201	290	240	260	260	230- 270	

200	149	240	183	206	210	160- 220	
150	98	180	126	155	160	90- 160	
100	46	110	68	106	110- 130	60- 100	
50	5 BC	70	11	51	70	40- 80	
0	57	60	47 BC	10 BC	50	20- 60	
50 BC	108	BC 100- 50 AD	106	70	BC 60- 10 AD	BC 140- 30 AD	
100	159	BC 130- 60 AD	164	140	120- 70 BC	180- 60 BC	
150	211	150 BC	224	179	190- 170	240- 170	
200	262	330- 170	283	257	380- 210	410- 200	
250	314	380- 200	343	310	390	420- 250	
300	365	400	403	389	410	430- 400	
350	417	420	463	473	430	450- 430	
400	468	460	524	528	460- 440	490- 450	430- 320 BC
450	520	770- 520	585	561	640- 500	590- 490	520- 420
500	571	780- 530	646	618	720- 660	780- 530	590- 510
550	623	780- 560	708	686	760	810- 560	660- 580
600	674	790	770	760	790	840- 790	720- 650
650	725	860	832	828	805	860- 820	780- 720
700	777	870	894	888	880- 850	890- 850	830- 770
750	828	880	956	937	900	950- 890	890- 830
800	880	980- 890	1019	979	990- 940	990- 950	940- 880
850	931	1070	1082	1042	1020	1020- 990	1000- 930
900	983	1120	1145	1116	1100	1150-1020	1050- 980
950	1034	1220-1140	1208	1183	1180-1160	1190-1150	1110-1030
1000	1086	1330-1230	1271	1256	1270-1240	1300-1200	1160-1080
1050	1137	1340	1335	1327	1300	1370-1260	1220-1130
1100	1188	1350	1398	1398	1390-1370	1410-1320	1270-1180
1150	1240	1460	1462	1455	1460	1510-1410	1330-1240
1200	1291	1490	1526	1497	1500	1520-1480	1380-1290
1250	1343	1510	1589	1556	1560-1520	1560-1520	1430-1340
1300	1394	1610-1530	1653	1626	1630-1600	1640-1560	1490-1400
1350	1446	1650	1717	1680	1680-1660	1670-1640	1560-1460
1400	1497	1670	1781	1745	1710-1690	1700-1650	1610-1510
1450	1549	1690	1846	1819	1870-1720	1880-1700	1680-1570
1500	1600	1720	1910	1873	1800	1940-1880	1740-1630

5568 C-14 date	5730 C-14 date	Suess	Wendland & Donley	Switsur	MASCA	50-year averaging	Egyptian historical
1550 BC	1652 BC	2040-1760 BC	1974 BC	1942 BC	2020-2000 BC	2060-1970 BC	1820-1700 BC
1600	1703	2070	2038	2031	2060	2090-2060	1830-1770
1650	1754	2090	2102	2094	2110	2110-2080	1970-1850
1700	1806	2120	2167	2147	2140	2140-2090	2040-1910
1750	1857	2140	2231	2199	2160	2180-2110	2180-2010
1800	1909	2160	2295	2267	2180	2210-2150	2210-2080
1850	1960	2180	2359	2342	2290-2190	2350-2180	2310-2170
1900	2012	2470-2210	2423	2355	2410-2340	2460-2220	2390-2250
1950	2063	2480-2410	2487	2427	2480	2520-2420	2500-2340
2000	2115	2490	2551	2535	2560	2550-2470	2570-2420
2050	2166	2510	2615	2651	2610	2580-2520	2670-2510
2100	2217	2740-2530	2679	2724	2780-2650	2830-2560	2740-2580
2150	2269	2930-2550	2742	2751	2830	2870-2610	2830-2650
2200	2320	2940	2808	2800	2900-2880	2900-2850	2890-2710
2250	2372	2940	2869	2891	2940-2920	2950-2870	2970-2770
2300	2423	2950	2932	2979	2970	3010-2910	3020-2830
2350	2475	2960	2995	3035	3110-2990	3150-2960	3070-2870
2400	2526	3340-2980	3058	3069	3150	3190-3110	3120-2920
2450	2578	3370-3210	3121	3119	3160	3240-3140	3210-3000
2500	2629	3380-3250	3184	3191	3310-3190	3330-3160	
2550	2681	3390	3246	3258	3330-3220	3370-3220	
2600	2732	3390	3308	3325	3370-3350	3390-3270	
2650	2783	3460-3400	3370	3381	3380	3430-3320	
2700	2835	3490-3400	3432	3425	3470-3400	3500-3350	
2750	2886	3510	3493	3458	3540-3500	3610-3410	
2800	2938	3640-3530	3554	3488	3600-3580	3670-3490	
2850	2989	3670	3615	3538	3630	3690-3640	
2900	3041	3680	3676	3601	3670	3710-3680	
2950	3092	3690	3736	3668	3700	3720-3690	
3000	3144	3710	3796	3724	3740	3740-3700	

3050	3195	3740	3856	3793	3770	3800-3710
3100	3246	3900-3820	3916	3880	3850-3800	3940-3770
3150	3298	3950	3975	3959	3900	3990-3910
3200	3349	3970	4033	4022	3980-3960	4020-3950
3250	3401	3990	4092	4080	4010	4050-3980
3300	3452	4220-4030	4150	4153	4120-4070	4140-4010
3350	3504	4330-4230	4207	4201	4190	4240-4040
3400	3555	4340	4265	4233	4330-4270	4480-4120
3450	3607	4350	4322	4261	4350-4330	4430-4240
3500	3658	4360	4378	4300	4380	(4460-4360)
3550	3710	4370	4434	4363	4410	(4500-4410)
3600	3761	4400	4490	4415	4440	(4540-4450)
3650	3812	4440	4545	4466	4470	(4570-4490)
3700	3864	4480	4599	4517	4500	(4620-4530)
3750	3915	4520	4654	4569	4560	(4660-4570)
3800	3967	4580	4707	4640	4590	(4700-4620)
3850	4018	4680	4761	4691	4640-4610	(4740-4660)
3900	4070	4810	4813	4736	4690-4670	4790-4710
3950	4121	4830	4865	4797	4870-4770	4890-4750
4000	4173	4860	4917	4843	4900	4930-4840
4050	4224	4890	4968	4891	4930	4960-4920
4100	4275	4940	5019	4955	5000	5010-4950
4150	4327	5000	5069	5008	5030	5070-5000
4200	4378	5070	5118	5058	5080	5160-5040
4250	4430	5220	5167	5096	5140-5110	5220-5110
4300	4481	5290	5216	5137	5240-5210	5260-5170
4350	4533		5263	5179	5250	5290-5220
4400	4584		5310	5212	5280	5340-5250
4450	4636		5357	5251	5310	5420-5270
4500	4590		5403	5298	5320	5480-5330

Appendix IB

Conversion of standard C-14 *dates to tree-ring calendar years*

NOTES

(a) All C-14 dates are standard 5568-year half-life figures.
(b) Results are based upon a fifty-year averaging procedure, i.e. C-14 dates within successive fifty-year tree-ring intervals were used to calculate means and appropriate standard errors over each fifty-year interval. Smooth curves through the upper and lower limits were drawn and maximum and minimum tree-ring dates derived for chosen values of C-14 date. The maximum and minimum figures have a 68 per cent probability of containing the true tree-ring date.
(c) The period between 3850 and 3500 BC in C-14 years is largely estimated, through lack of data.

Table 1. 1800–910 AD.

5568 C-14 date	Tree-ring range	5568 C-14 date	Tree-ring range	5568 C-14 date	Tree-ring range
1800 AD	1650-1760 AD	1500 AD	1390-1430 AD	1200 AD	1230-1260 AD
1790	1650-1740	1490	1390-1430	1190	1220-1250
1780	1650-1720	1480	1380-1420	1180	1210-1250
1770	1640-1700	1470	1380-1420	1170	1210-1240
1760	1640-1680	1460	1370-1410	1160	1200-1240
1750	1640-1660	1450	1370-1410	1150	1190-1230
1740	1630-1660	1440	1370-1410	1140	1170-1220
1730	1610-1660	1430	1360-1400	1130	1150-1220
1720	1600-1650	1420	1360-1400	1120	1120-1210
1710	1580-1650	1410	1350-1390	1110	1100-1210
1700 AD	1570-1650 AD	1400 AD	1350-1390 AD	1100 AD	1080-1200 AD
1690	1550-1640	1390	1340-1390	1090	1070-1170
1680	1540-1630	1380	1340-1380	1080	1070-1150
1670	1520-1610	1370	1330-1380	1070	1060-1120
1660	1510-1600	1360	1330-1370	1060	1060-1100
1650	1490-1590	1350	1320-1370	1050	1050-1070
1640	1480-1570	1340	1310-1360	1040	1040-1060
1630	1470-1560	1330	1310-1360	1030	1030-1060
1620	1470-1540	1320	1300-1350	1020	1030-1050
1610	1460-1530	1310	1300-1350	1010	1020-1050
1600 AD	1450-1510 AD	1300 AD	1290-1340 AD	1000 AD	1010-1040 AD
1590	1440-1500	1290	1280-1330	990	1000-1030
1580	1430-1490	1280	1280-1320	980	990-1030
1570	1430-1480	1270	1270-1310	970	980-1020
1560	1420-1470	1260	1270-1300	960	970-1020
1550	1410-1460	1250	1260-1290	950	960-1010
1540	1410-1450	1240	1250-1280	940	950-1000
1530	1400-1450	1230	1250-1280	930	940-990
1520	1400-1440	1220	1240-1270	920	940-980
1510	1390-1440	1210	1240-1270	910	930-970

Table 2. 900–10 AD.

5568 C-14 date	Tree-ring range	5568 C-14 date	Tree-ring range	5568 C-14 date	Tree-ring range
900 AD	920-960 AD	600 AD	610-640 AD	300 AD	300-350 AD
890	910-950	590	600-640	290	290-330
880	900-940	580	600-630	280	270-320
870	880-930	570	590-630	270	260-300
860	870-920	560	590-620	260	240-290
850	860-910	550	580-620	250	230-270
840	840-890	540	570-610	240	220-260
830	830-880	530	560-600	230	200-250
820	810-860	520	550-600	220	190-240
810	800-850	510	540-590	210	170-230
800 AD	780-830 AD	500 AD	530-580 AD	200 AD	160-220 AD
790	770-820	490	520-570	190	150-210
780	770-810	480	510-550	180	130-200
770	760-810	470	490-540	170	120-180
760	760-800	460	480-520	160	100-170
750	750-790	450	470-510	150	90-160
740	730-780	440	460-500	140	80-150
730	720-770	430	450-490	130	80-140
720	700-760	420	430-470	120	70-120
710	690-750	410	420-460	110	70-110
700 AD	670-740 AD	400 AD	410-450 AD	100 AD	60-100 AD
690	660-720	390	400-440	90	60-100
680	650-710	380	400-430	80	50-90
670	650-690	370	390-430	70	50-90
660	640-680	360	390-420	60	40-80
650	630-660	350	380-410	50	40-80
640	630-660	340	360-400	40	40-80
630	620-650	330	350-390	30	30-70
620	620-650	320	330-370	20	30-70
610	610-640	310	320-360	10	20-60

Table 3. 0–890 BC.

5568 C-14 date	Tree-ring range	5568 C-14 date	Tree-ring range	5568 C-14 date	Tree-ring range
0 BC	20-60 AD	300 BC	430-400 BC	600 BC	840-790 BC
10	BC 10-50	310	430-410	610	840-800
20	40-40	320	440-410	620	850-800
30	80-40	330	440-420	630	850-810
40	110-30	340	450-420	640	860-810
50	140-30	350	450-430	650	860-820
60	150-10	360	460-430	660	870-830
70	160-10 BC	370	470-440	670	870-830
80	160-20	380	470-440	680	880-840
90	170-40	390	480-450	690	880-840
100 BC	180-60 BC	400 BC	490-450 BC	700 BC	890-850 BC
110	190-80	410	510-460	710	900-860
120	200-100	420	530-470	720	910-870
130	220-130	430	550-470	730	930-870
140	230-150	440	570-480	740	940-880
150	240-170	450	590-490	750	950-890
160	270-180	460	630-500	760	960-900
170	310-180	470	670-510	770	970-910
180	340-190	480	700-510	780	970-930
190	380-190	490	740-520	790	980-940
200 BC	410-200 BC	500 BC	780-530 BC	800 BC	990-950 BC
210	410-210	510	790-540	810	1000-960
220	410-220	520	790-540	820	1000-970
230	420-230	530	800-550	830	1010-970
240	420-240	540	800-550	840	1010-980
250	420-250	550	810-560	850	1020-990
260	420-280	560	820-610	860	1050-1000
270	420-310	570	820-650	870	1070-1000
280	430-340	580	830-700	880	1100-1010
290	430-370	590	830-740	890	1120-1010

Appendix IB. C-14 Dates to Tree-ring Years

Table 4. 900–1790 BC.

5568 C-14 date	Tree-ring range	5568 C-14 date	Tree-ring range	5568 C-14 date	Tree-ring range
900 BC	1150-1020 BC	1200 BC	1520-1480 BC	1500 BC	1940-1880 BC
910	1160-1050	1210	1530-1490	1510	1960-1900
920	1170-1070	1220	1540-1500	1520	1990-1920
930	1170-1100	1230	1540-1500	1530	2010-1930
940	1180-1120	1240	1550-1510	1540	2040-1950
950	1190-1150	1250	1560-1520	1550	2060-1970
960	1210-1160	1260	1580-1530	1560	2070-1990
970	1230-1170	1270	1590-1540	1570	2070-2010
980	1260-1180	1280	1610-1540	1580	2080-2020
990	1280-1190	1290	1620-1550	1590	2080-2040
1000 BC	1300-1200 BC	1300 BC	1640-1560 BC	1600 BC	2090-2060 BC
1010	1310-1210	1310	1650-1580	1610	2090-2060
1020	1330-1220	1320	1650-1590	1620	2100-2070
1030	1340-1240	1330	1660-1610	1630	2100-2070
1040	1360-1250	1340	1660-1620	1640	2110-2080
1050	1370-1260	1350	1670-1640	1650	2110-2080
1060	1380-1270	1360	1680-1640	1660	2120-2080
1070	1390-1280	1370	1680-1640	1670	2120-2080
1080	1390-1300	1380	1690-1650	1680	2130-2090
1090	1400-1310	1390	1690-1650	1690	2130-2090
1100 BC	1410-1320 BC	1400 BC	1700-1650 BC	1700 BC	2140-2090 BC
1110	1430-1340	1410	1740-1660	1710	2150-2090
1120	1450-1360	1420	1770-1670	1720	2160-2100
1130	1470-1370	1430	1810-1680	1730	2160-2100
1140	1490-1390	1440	1840-1690	1740	2170-2110
1150	1510-1410	1450	1880-1700	1750	2180-2110
1160	1510-1420	1460	1890-1740	1760	2190-2120
1170	1510-1440	1470	1900-1770	1770	2190-2130
1180	1520-1450	1480	1920-1810	1780	2200-2130
1190	1520-1470	1490	1930-1840	1790	2200-2140

Table 5. 1800–2690 BC.

5568 C-14 date	Tree-ring range	5568 C-14 date	Tree-ring range	5568 C-14 date	Tree-ring range
1800 BC	2210-2150 BC	2100 BC	2830-2560 BC	2400 BC	3190-3110 BC
1810	2240-2160	2110	2840-2570	2410	3200-3120
1820	2270-2160	2120	2850-2580	2420	3210-3120
1830	2290-2170	2130	2850-2590	2430	3220-3130
1840	2320-2170	2140	2860-2600	2440	3230-3130
1850	2350-2180	2150	2870-2610	2450	3240-3140
1860	2370-2190	2160	2880-2660	2460	3260-3140
1870	2390-2200	2170	2880-2710	2470	3280-3150
1880	2420-2200	2180	2890-2750	2480	3290-3150
1890	2440-2210	2190	2890-2800	2490	3310-3160
1900 BC	2460-2220 BC	2200 BC	2900-2850 BC	2500 BC	3330-3160 BC
1910	2470-2260	2210	2910-2850	2510	3340-3170
1920	2480-2300	2220	2920-2860	2520	3350-3180
1930	2500-2340	2230	2930-2860	2530	3350-3200
1940	2510-2380	2240	2940-2870	2540	3360-3210
1950	2520-2420	2250	2950-2870	2550	3370-3220
1960	2530-2430	2260	2960-2880	2560	3370-3230
1970	2530-2440	2270	2970-2890	2570	3380-3240
1980	2540-2450	2280	2990-2890	2580	3380-3250
1990	2540-2460	2290	3000-2900	2590	3390-3260
2000 BC	2550-2470 BC	2300 BC	3010-2910 BC	2600 BC	3390-3270 BC
2010	2560-2480	2310	3040-2920	2610	3400-3280
2020	2560-2490	2320	3070-2930	2620	3410-3290
2030	2570-2500	2330	3090-2940	2630	3410-3300
2040	2570-2510	2340	3120-2950	2640	3420-3310
2050	2580-2520	2350	3150-2960	2650	3430-3320
2060	2630-2530	2360	3160-2990	2660	3440-3330
2070	2680-2540	2370	3170-3020	2670	3460-3330
2080	2730-2540	2380	3170-3050	2680	3470-3340
2090	2780-2550	2390	3180-3080	2690	3490-3340

Table 6. 2700–3590 BC.

5568 C-14 date	Tree-ring range	5568 C-14 date	Tree-ring range	5568 C-14 date	Tree-ring range
2700 BC	3500-3350 BC	3000 BC	3740-3700 BC	3300 BC	4140-4010 BC
2710	3520-3360	3010	3750-3700	3310	4160-4020
2720	3540-3370	3020	3760-3700	3320	4180-4020
2730	3570-3390	3030	3780-3710	3330	4200-4030
2740	3590-3400	3040	3790-3710	3340	4220-4030
2750	3610-3410	3050	3800-3710	3350	4240-4040
2760	3620-3430	3060	3830-3720	3360	4290-4060
2770	3630-3440	3070	3860-3730	3370	4340-4070
2780	3650-3460	3080	3880-3750	3380	4380-4090
2790	3660-3470	3090	3910-3760	3390	4430-4100
2800 BC	3670-3490 BC	3100 BC	3940-3770 BC	3400 BC	4480-4120 BC
2810	3670-3520	3110	3950-3800	3410	4470-4140
2820	3680-3550	3120	3960-3830	3420	4460-4170
2830	3680-3580	3130	3970-3850	3430	4450-4190
2840	3690-3610	3140	3980-3880	3440	4440-4220
2850	3690-3640	3150	3990-3910	3450	4430-4240
2860	3690-3650	3160	4000-3920	3460	4440-4260
2870	3700-3660	3170	4000-3930	3470	4440-4290
2880	3700-3660	3180	4010-3930	3480	4450-4310
2890	3710-3670	3190	4010-3940	3490	4450-4340
2900 BC	3710-3680 BC	3200 BC	4020-3950 BC	3500 BC	4460-4360 BC
2910	3710-3680	3210	4030-3960	3510	4470-4370
2920	3710-3680	3220	4030-3960	3520	4480-4380
2930	3720-3690	3230	4040-3970	3530	4480-4390
2940	3720-3690	3240	4040-3970	3540	4490-4400
2950	3720-3690	3250	4050-3980	3550	4500-4410
2960	3720-3690	3260	4070-3990	3560	4510-4420
2970	3730-3690	3270	4090-3990	3570	4520-4430
2980	3730-3700	3280	4100-4000	3580	4520-4430
2990	3740-3700	3290	4120-4000	3590	4530-4440

Table 7. 3600–4500 BC.

5568 C-14 date	Tree-ring range	5568 C-14 date	Tree-ring range	5568 C-14 date	Tree-ring range
3600 BC	4540-4450 BC	3900 BC	4790-4710 BC	4200 BC	5160-5040 BC
3610	4550-4460	3910	4810-4720	4210	5170-5050
3620	4550-4470	3920	4830-4730	4220	5180-5070
3630	4560-4470	3930	4850-4730	4230	5200-5080
3640	4560-4480	3940	4870-4740	4240	5210-5100
3650	4570-4490	3950	4890-4750	4250	5220-5110
3660	4580-4500	3960	4900-4770	4260	5230-5120
3670	4590-4510	3970	4910-4790	4270	5240-5130
3680	4600-4510	3980	4910-4800	4280	5240-5150
3690	4610-4520	3990	4920-4820	4290	5250-5160
3700 BC	4620-4530 BC	4000 BC	4930-4840 BC	4300 BC	5260-5170 BC
3710	4630-4540	4010	4940-4860	4310	5270-5180
3720	4640-4550	4020	4940-4870	4320	5270-5190
3730	4640-4550	4030	4950-4890	4330	5280-5200
3740	4650-4560	4040	4950-4900	4340	5280-5210
3750	4660-4570	4050	4960-4920	4350	5290-5220
3760	4670-4580	4060	4970-4930	4360	5300-5230
3770	4680-4590	4070	4980-4930	4370	5310-5230
3780	4680-4600	4080	4990-4940	4380	5320-5240
3790	4690-4610	4090	5000-4940	4390	5330-5240
3800 BC	4700-4620 BC	4100 BC	5010-4950 BC	4400 BC	5340-5250 BC
3810	4710-4630	4110	5020-4960	4410	5360-5250
3820	4720-4640	4120	5030-4970	4420	5370-5260
3830	4720-4640	4130	5050-4980	4430	5390-5260
3840	4730-4650	4140	5060-4990	4440	5400-5270
3850	4740-4660	4150	5070-5000	4450	5420-5270
3860	4750-4670	4160	5090-5010	4460	5430-5280
3870	4760-4680	4170	5110-5020	4470	5440-5290
3880	4770-4690	4180	5120-5020	4480	5460-5310
3890	4780-4700	4190	5140-5030	4490	5470-5320
				4500	5480-5330

Appendix IC

Conversion of standard C-14 *dates to calendar years, based on the Egyptian historical curve*

NOTES

(a) All C-14 dates are standard 5568-year half-life figures.

(b) Results are based upon a polynomial regression procedure for C-14 dates from the Egyptian historical series of the BM/UCLA programme. The correction range limits are derived from both the carbon dating standard deviations and the relevant historical accuracy.

(c) For C-14 dates between 2450 and 1800 BC there is good agreement with tree-ring correction figures. After 1800 BC agreement becomes poorer.

Table 1. 600–1490 BC.

5568 C-14 date	Historical curve range	5568 C-14 date	Historical curve range	5568 C-14 date	Historical curve range
600 BC	720-650 BC	700 BC	830-770 BC	800 BC	940-880 BC
610	730-660	710	840-780	810	950-890
620	740-680	720	850-790	820	960-900
630	760-690	730	870-810	830	980-910
640	770-710	740	880-820	840	990-920
650	780-720	750	890-830	850	1000-930
660	790-730	760	900-840	860	1010-940
670	800-740	770	910-850	870	1020-950
680	810-750	780	920-860	880	1030-960
690	820-760	790	930-870	890	1040-970
900 BC	1050-980 BC	1000 BC	1160-1080 BC	1100 BC	1270-1180 BC
910	1060-990	1010	1170-1090	1110	1280-1190
920	1070-1000	1020	1180-1100	1120	1290-1200
930	1090-1010	1030	1200-1110	1130	1310-1220
940	1100-1020	1040	1210-1120	1140	1320-1230
950	1110-1030	1050	1220-1130	1150	1330-1240
960	1120 1040	1060	1230 1140	1160	1340-1250
970	1130-1050	1070	1240-1150	1170	1350-1260
980	1140-1060	1080	1250-1160	1180	1360-1270
990	1150-1070	1090	1260-1170	1190	1370-1280
1200 BC	1380-1290 BC	1300 BC	1490-1400 BC	1400 BC	1610-1510 BC
1210	1390-1300	1310	1500-1410	1410	1620-1520
1220	1400-1310	1320	1520-1420	1420	1640-1530
1230	1410-1320	1330	1530-1440	1430	1650-1550
1240	1420-1330	1340	1550-1450	1440	1670-1560
1250	1430-1340	1350	1560-1460	1450	1680-1570
1260	1440-1350	1360	1570-1470	1460	1690-1580
1270	1450-1360	1370	1580-1480	1470	1710-1590
1280	1470-1380	1380	1590-1490	1480	1720-1610
1290	1480-1390	1390	1600-1500	1490	1730-1620

Table 2. 1500–2390 BC.

5568 C-14 date	Historical curve range	5568 C-14 date	Historical curve range	5568 C-14 date	Historical curve range
1500 BC	1740-1630 BC	1600 BC	1880-1770 BC	1700 BC	2040-1910 BC
1510	1760-1640	1610	1900-1790	1710	2060-1930
1520	1770-1660	1620	1920-1800	1720	2080-1950
1530	1790-1670	1630	1930-1820	1730	2090-1970
1540	1800-1690	1640	1950-1830	1740	2110-1990
1550	1820-1700	1650	1970-1850	1750	2130-2010
1560	1830-1710	1660	1980-1860	1760	2150-2020
1570	1840-1730	1670	2000-1870	1770	2160-2040
1580	1860-1740	1680	2010-1890	1780	2180-2050
1590	1870-1760	1690	2030-1900	1790	2190-2070
1800 BC	2210-2080 BC	1900 BC	2390-2250 BC	2000 BC	2570-2420 BC
1810	2230-2100	1910	2410-2270	2010	2590-2440
1820	2250-2120	1920	2430-2290	2020	2610-2460
1830	2270-2130	1930	2460-2300	2030	2630-2470
1840	2290-2150	1940	2480-2320	2040	2650-2490
1850	2310-2170	1950	2500-2340	2050	2670-2510
1860	2330-2190	1960	2510-2360	2060	2680-2520
1870	2340-2200	1970	2530-2370	2070	2700-2540
1880	2360-2220	1980	2540-2390	2080	2710-2550
1890	2370-2230	1990	2560-2400	2090	2730-2570
2100 BC	2740-2580 BC	2200 BC	2890-2710 BC	2300 BC	3020-2830 BC
2110	2760-2590	2210	2910-2720	2310	3030-2840
2120	2780-2610	2220	2920-2730	2320	3040-2850
2130	2790-2620	2230	2940-2750	2330	3050-2850
2140	2810-2640	2240	2950-2760	2340	3060-2860
2150	2830-2650	2250	2970-2770	2350	3070-2870
2160	2840-2660	2260	2980-2780	2360	3080-2880
2170	2850-2670	2270	2990-2790	2370	3090-2890
2180	2870-2680	2280	3000-2810	2380	3100-2900
2190	2880-2690	2290	3010-2820	2390	3110-2910

Appendix II

The Role of the Archaeologist in C-14 Age Measurement

D. D. HARKNESS

A C-14 date may be considered to exist in two forms: (a) the 'conventional age' as reported by C-14 laboratories and/or published in the journal *Radiocarbon*, and (b) the so-called 'corrected age' derived after recourse to half-life changes and calibration curves or tables. While a great deal of advice has been presented to the archaeologist on how he should best correct and interpret the basic conventional C-14 date the importance of his role in ensuring its validity has been largely neglected.

This article would hope to clarify where and why the archaeologist can assist in the practical aspects of dating and in so doing also answer such familiar questions as: Which sample material is most suited to the production of a valid C-14 date?; How much material is required?; How are samples best collected, recorded, stored and transported?

Conventional C-14 Dates

It is perhaps necessary to first of all establish the conventional C-14 date in its true context. To do so we must appreciate that C-14 dating is a purely physical technique whereby the residual radioactivity in a sample is determined within known limits of accuracy. Subsequent conversion of this measurement to a range of 'C-14 years', in the familiar form AGE±ERROR BP, involves somewhat complex mathematical and statistical treatments (Callow *et al.* 1965) which are, however, rigidly followed by C-14 laboratories throughout the world. Thus all *conventional dates* should be directly comparable, irrespective of the method of C-14 assay.

The price which must be paid for this desirable uniformity involves adherence to the basic principles of the C-14 dating theory. Thus in calculation of a conventional age it is assumed that (a) the radioactive half-life for C-14 is accurate, and (b) the ratio of C-14 to C-12 atoms in living materials has remained universally constant throughout the time span of the chronology. With the improvement in scientific methods, and current comparisons of C-14 ages with both alternative physio/chemical dating methods and historical documentation, it has become evident that these basic assumptions are not strictly valid. For example, it is now generally accepted that the half-life value used in age calculation is some three per cent too short (Godwin 1962), and the implications of calibration data provided by dendrochronological and historical records are an all too familiar source of controversy. These deviations from the basic theory are of course not reflected in the conventional C-14 date and hence are outwith the purpose of this discussion.

Suffice it to say, therefore, that although conventional C-14 dates are founded on an oversimplified picture of past geochemical and geophysical conditions this fact does not invalidate the C-14 dating method, but in the light of present knowledge only limits its application.

Sources of Error

Since W.F. Libby first proposed the method some twenty-five years ago, several thousand C-14 ages have been measured and reported. While the vast majority of these have been accepted as providing an accurate estimate of the age of the events they proport to date, several are clearly anomalous even outwith the limitations discussed above. In such cases the identification, collection or measurement of the sample material must be suspect. For the most part, avoidance of such errors can only be achieved by close and continual collaboration between the person responsible for collection of dateable materials and the C-14 chronologist to whom these samples are eventually submitted. The fundamental reasons are obvious. Few laboratory personnel have more than a scant theoretical knowledge of the problems encountered in archaeological prospection, nor is it reasonable to expect the archaeologist to be fully conversant with the complex scientific procedures involved in C-14 assay.

It seems worthwhile to review both the possible sources of anomalous age measurements and how, where possible, these can be avoided.

Statistical limitations. By definition a C-14 age is never an absolute number, but a range of years within which there is a definite probability that the true radiometric date falls. Usually results are quoted at the $\pm\sigma$ confidence level, i.e. there is a 68 per cent probability that the true age falls somewhere within the limits defined by the 'error' term. Doubling the quoted range ($\pm 2\sigma$) increases the probability to 95 per cent, but no broadening of the limits can ever give one hundred per cent certainty. It is perhaps then a sobering thought that, for dates quoted as per *Radiocarbon*, one in three does not bracket the true radiometric age, and even after doubling the uncertainty range one in twenty remains unavoidably anomalous. Clearly, controversial arguments should not be based on a single C-14 date no matter how meticulously the sample may have been collected and measured.

The C-14 *concentration has been wrongly determined.* No laboratory can claim to be infallible, and the possibility certainly exists. Every reputable laboratory has however a built-in system of cross-checking whereby sample identification, chemical procedures and equipment performance is constantly monitored and detailed records kept. It is also the policy at SURRC to exchange samples with other established laboratories as an overall check on the total dating process.

Where possible, sufficient sample should be collected to allow a duplicate analysis in the rare event that this should prove necessary as a result of equipment malfunction or operator error.

The sample does not correspond exactly with the event to be dated. This type of error is more common than most collectors realise or perhaps care to admit. Purely human errors such as mislabelling or mistaken identity in the laboratory, etc., come into this category but these can usually be traced and remedied. The more fundamental type of mis-association is probably best described by considering some hypothetical but typical situations.

If we consider charcoal excavated from a prehistoric hearth the obvious assumption is that this material is contemporaneous with occupation of the site. The assumption is reasonable if the wood burned came from the young branches of trees, however if there is a possibility that the fuel used was driftwood or the heartwood of long-lived trees then the date obtained would tend to give only a maximum possible age. Should fossil wood or peat have been used then the date would be totally misleading. Similarly wood samples from massive structures should be taken from the outermost surfaces, i.e. the youngest growth rings.

In the same context consider the use of shells to date an occupation level. If these are relics of a food source then they will indeed date the consumers. If however they represent artifacts or ornaments they are likely to have been older shells selected from a beach deposit in which case the radiometric age is again only of the maximum possible type.

In terms of stratification it should also be considered whether the sample has remained *in situ* or has possibly been disturbed or deposited by such agencies as burrowing animals, natural erosion processes or later human activity. Clearly these observations can only be made by the collector.

The major point of emphasis is that C-14 dating can only indicate when the fossil organic material lived, not when it was employed in prehistory. Only the archaeologist, by careful perception during excavation and some prior knowledge of the people he wishes to date, can avoid pitfalls of the type described.

The sample has been contaminated prior or subsequent to its collection. This is undoubtedly the most familiar source of misleading ages. Contaminants in this sense are carbonaceous materials which are not contemporaneous with the original living sample. If not removed prior to C-14 measurement these foreign materials result in an increased or reduced C-14 activity (tables 1 and 2).

Contamination can result from natural processes or can be introduced during collection or storage, and it is a rare sample indeed that does not contain some measurable quantity of natural contaminant. Fortunately, laboratory procedures can be designed to totally remove foreign carbon but again the assistance of the field worker in recognising their existence and composition is essential.

Table 1. Effect of contamination by 'infinitely old' carbon on the true radiometric age.

Contamination %	Years older than true age
5	400
10	830
20	1 800
30	2650
40	4100

Table 2. Effect of contamination by modern (pre-bomb) carbon on the true radiometric age.

True age (years)	Apparent age (years) obtained as result of: 1% contamination	5% contamination	10% contamination
600	540	160	modern
1000	910	545	160
5000	4870	4230	3630
10000	9730	8710	7620
infinitely old	36600	24000	18400

Natural contaminants include younger rootlets in peat or soil samples, water-soluble humic acids absorbed in charcoal or wood, exchanged carbonate in the mineral structure of bone or in the outer layers of shell. These sources of error may not always be readily obvious in the field, but if comprehensive notes are made on the stratigraphy and local site environment then laboratory workers have a firm basis for pre-treatment of the particular sample prior to C-14 measurement.

Artificial contamination is more readily recognised in the laboratory but of course every attempt should be made to avoid its occurrence. Some typical examples here are paint or grease from tools, hair from brushes, unsuitable packaging material, mould growth due to lengthy storage in a damp state. Organically based preservatives must be avoided at all costs. Again the onus is on the collector to ensure that practices which could result in such inclusions are avoided or, where their possible existence occurs, to draw the attention of the dating laboratory to this fact.

Selection of Samples

In the selection of sample material for dating the first criterion is obviously to minimise the possibility of uncertain or erroneous age association as discussed previously. Where a choice of dateable material still remains, selection should then give preference to considerations of (a) the size of sample available and (b) its susceptibility to the more obscure forms of contamination. Typical materials are best discussed individually.

Wood, whatever its state of preservation, is an ideal dating material. It is possible to isolate and purify the cellulose component, and by dating only this fraction all possible contaminants are avoided. Approximately 40 per cent of the weight of dry, well-preserved wood is recoverable as cellulose, but it must be borne in mind that up to 90 per cent of the total weight of waterlogged samples may be due to the moisture content.

Charcoal, like wood cellulose, is chemically very stable and stringent pre-treatment can be applied without undue loss of the dateable component. However, it is an excellent sponge for humic acids and other water-soluble contaminants, and while these can be completely removed in the laboratory they often constitute more than 50 per cent of the original sample weight. When the charcoal sample exists as small flakes dispersed in a soil matrix, insufficient sample is invariably submitted to allow for the necessary decon-tamination procedures. It is useful to bear in mind that, volume for volume, charcoal is about four times lighter than soil or sand.

Peat is also an excellent material for dating. It is true that there can be vertical transport of soluble humic components through the peat profile, but these are readily removed by laboratory pre-treatment leaving the plant material which formed *in situ*. There is also a potential source of error from roots of a later generation of plants growing through the already formed peat, but again these are readily recognised in most cases and can be eliminated.

Bone and Antler. Doubts have been cast in some publications about the validity of C-14 dates produced from these materials. It is our experience at SURRC that the problem of carbon exchange after death does not occur for the protein (collagen) fraction. Consequently, extraction of the bone protein and its further purification to gelatin ensures that no contaminants remain in the dated component. The amount of gelatin which can be recovered from bones is unfortunately very variable, and depends on their state of preservation. In ideal cases as much as 25 per cent of the original weight of bone is usable, but samples which yield little or no collagen are common.

Whalebone should be avoided. It is suspected that these animals can derive their food supply from regions of the ocean subject to upwelling of deep, and hence old, water. The primary source of carbon in the food chain (dissolved bicarbonate) can have an apparent C-14 age of up to 1 000 years. This would be reflected to some indeterminable degree in the whale tissue.
Shells, both marine and terrestrial, can, within certain limitations, provide useful C-14 dates. They should however be considered as least suited to precise dating, and hence only considered where alternative material is absent or as a secondary check to such material.

The problems involved in shell dating relate to both geochemical uncertainty as to their true $^{14}C/^{12}C$ content at time of death, and the ease with which the dead carbonate structure can exchange with carbon dioxide dissolved in groundwaters. Probably those shells whose habitat is an open beach remote from outcrops of chalk or limestone are most satisfactory when shell dating must be envisaged. In all cases, however, each proposed sample submission should first be discussed with the C-14 laboratory involved.
Lake mud, soil, turf, etc. These materials are less familiar in the archaeological context but they can very often provide reliable C-14 ages. Generally they are susceptible to the more obscure forms of natural contamination, and therefore their suitability in relation to a particular site is very dependent on the local environmental conditions. Again, it is recommended that the advice of the C-14 laboratory be sought prior to the proposed application of these materials for age determination.

It must be emphasised that no sample should be submitted for dating without full details of its nature, stratigraphy, environment etc. Most C-14 laboratories provide sample description forms designed to assist in recording the relevant information. Thoughtful completion of these reports should be regarded as an extremely important aspect of the overall dating effort.

Sample Size

It is very important that an adequate sample be collected and submitted. For actual measurement most C-14 laboratories, irrespective of their equipment sophistication, require a minimum of 1 gm decontaminated carbon. The optimum amount for liquid scintillation counting, as applied at SURRC, is 15 gm (Harkness and Wilson 1972).

The all-important 'error term' associated with every C-14 date is a function of sample age but also, much more significantly, of sample size. In order to gain a predetermined dating precision the relationship between sample size and counting time is exponential and for smaller samples the desired precision is often unobtainable. It is perhaps worthwhile to remember that several C-14 laboratories base their dating charge for individual samples on equipment utilisation, i.e. primarily on sample size.

Obviously the collector cannot be expected to assess exactly the amount of usable carbon he is submitting. As indicated above, appropriate amounts differ from one type of material to another and it is impossible to give anything other than a very general estimate of how much genuine sample is likely to be recovered after the decontamination processes. Table 3 attempts to give some figures for guidance, and optimum sample size should be the aim in all collections. No C-14 laboratory will complain about receipt of too much sample.

Table 3. Amounts of material required for C-14 dating.

	Approximate dry weight to be submitted		
Sample	minimum (gm)	optimum (gm)	Comments
wood	3	100	
charcoal	2	50	Only if relatively soil free.
peat	3	50	
bone or antler	10	500	For well-preserved material.
shell	15	500	Only whole shells should be considered.
lake mud (gyttja)	10	100	
turf	5	50	
soil	?	?	Estimate should be sought from C-14 lab, preferably on basis of small trial sample.

1 ounce~28 gm.

Sample Containment

Experience has shown that samples are best stored and transported in air-tight polythene bags or in clean plastic, glass or metal containers. Each sample should be individually packed, to avoid cross-contamination, and clearly labelled.

At SURRC we favour the use of heavy grade polythene bags, the sample being placed and sealed in one bag and this then inserted into a second, also firmly sealed. A card or slip of paper bearing the unambiguous sample identification, written in indelible ink, should be inserted between the inner and outer bags. Writing or tacky labels on the outer surface of containers should not be used since these have a tendency to rub off. Types of packing to avoid are cotton wool, sawdust, wood shavings and paper, since these constitute sources of artificial contamination as defined above. No treatment need be applied to the samples between their collection and delivery to the laboratory, indeed even simply oven drying can often mask evidence of contamination. The period between collection and submission for dating should be as short as possible, ideally less than one month.

Publication of Results

In reporting C-14 dates to the submitter the laboratory will normally assign a code number, e.g. SRR-101, to each result. This notation uniquely defines both the particular sample measured and the laboratory involved. The publication of C-14 dates without this internationally recognised form of reference is as irresponsible as it is confusing. Similarly no C-14 age should be quoted without its attendant error; the common practice of dropping this term in conversion from BP (before present) to AD/BC notation is to be discouraged. Where conventional ages are 'corrected' for interpretative purposes this should be clearly stated. Ideally the conventional age in its *Radiocarbon* format should be quoted simultaneously. This point is particularly important in view of the increasing usage in some archaeological journals of lower-case letters (ad/bc) to denote conventional C-14 ages. The upper case (AD/BC) has always been used in *Radiocarbon*, and presumably this will continue to be so. Unless extreme care is taken in publication the potential for utter confusion is obvious.

Although the results published in *Radiocarbon* appear with only brief details as to location and significance, this publication constitutes a very convenient international catalogue which ensures that a collector's work is made known to others. Hence the importance of the laboratory code numbers. It is in the archaeologist's interest, therefore, to ensure that his work is adequately recorded with bibliographic details of fuller discussions.

No dates are included in the SURRC lists to *Radiocarbon* without the prior permission of the submitter and subsequent notification of the entry format.

Appendix III
Handlist of Radiocarbon Laboratories

TREVOR WATKINS & DOUGLAS HARKNESS

This handlist is intended to be useful to active excavators anywhere who want to know where they may most appropriately seek to have samples assayed for C-14 dating. If it encourages archaeologists to think of having samples dated who otherwise may have been deterred by ignorance of cost or application procedures then so much the better: and better still, perhaps, if it encourages archaeologists planning excavations to allow some money in their estimates for the dating of samples recovered during excavation.

In the list appear those laboratories that are known to be active, which are also known to be open to archaeologists (at least in certain circumstances), and which replied to our enquiries. We make no claim to have produced an exhaustive list; that can be found in the journal *Radiocarbon*. Our purpose has been to provide a select list which gives basic information to archaeologists of a kind not readily available elsewhere. Any laboratory on the list provides at least some service to archaeologists, whereas a number on the general *Radiocarbon* list are available only for specific, non-archaeological purposes such as palaeobotany, oceanography or geology, or indeed are quite inactive. All laboratories on this list are able to do at least some archaeological dating work for archaeologists who have no connection with the laboratory or its parent institution, whereas a number of laboratories listed in *Radiocarbon* are available only to a certain restricted circle of users. Our list gives some indication to would-be users of the ability or willingness of laboratories to undertake archaeological dating work for the excavator who has no claim on any particular laboratory.

Secondly, the list which follows gives a statement from each laboratory concerning the kind of specialisation which it may pursue or special services it may offer. It may be of benefit to users to be able to choose a laboratory which gives special consideration to the kind of samples which they have to submit, or a laboratory which is specially concerned with the period or area from which the samples come. It may even be the case that the archaeologist can assist a laboratory by providing samples of particular usefulness to that laboratory's own research.

Two other pieces of information are given for each laboratory. It may be of concern to the archaeologist to know how long it is likely to take to obtain a C-14 date. On occasions archaeologists are impatient and want a date quickly, for instance, to publish with the preliminary note of an important discovery. Some laboratories can offer a quick service, while others may warn of a long wait. In any case it will help the excavator in planning and timing his publication to have an indication of how long he may have to wait for C-14 results.

Finally we have asked laboratories to indicate the scale of charges which they make for their services. The figures given were quoted by the laboratories themselves in the Spring of 1975 as being likely to be correct at the time of publication. With the passage of a short time, doubtless these figures will become less and less useful as accurate gauges of cost, but at least for a while they will be of use, and they will continue to provide a relative picture of the kind of differences between laboratories.

The laboratories are listed in the order in which they appear in *Radiocarbon*, that is in the alphabetical order of their recognised code-signatures.

ANU AUSTRALIAN NATIONAL UNIVERSITY

Address. Australian National University Radiocarbon Dating Laboratory, Research School of Earth Sciences, Box 4, Canberra, Australia 2600.

Availability. Priority given to work generated by ANU research staff, and to samples sponsored by the Australian Institute of Aboriginal Studies. Non-Australian samples must be related to research work being undertaken at ANU.

Special interests. Specialisations have been developed in dating soil organic matter and soil carbonate. Interest in the specialised techniques applicable to low-level liquid scintillation counting of C-14.

Turnaround time. Normally about 200 dates a year are produced in two reports at half-year intervals. In special circumstances a date may be produced in a week.

Charge per sample. Not a commercial dating laboratory, but a limited number of samples at Australian $150 seems unavoidable.

Birm BIRMINGHAM UNIVERSITY RADIOCARBON LABORATORY

Address. Prof. A. Williams, FRS, Department of Geological Sciences, University of Birmingham, PO Box 363, Birmingham B15 2TT, UK.

Availability. Generally available for archaeological dating, subject to charges and the right to refuse samples considered unsuitable. The laboratory expects to publish all results in *Radiocarbon*. In all cases archaeologists should consult the laboratory about a sample before submission.

Special interests. All dates are corrected for isotopic fractionation by measurement of $\delta^{13}C$, which is essential for common archaeological sample material such as grain, shell and bone. Bone dates are obtained from extracted collagen, not carbono-phosphate.
Turnaround time. Average is 12 weeks, but varying from 7 to 16 weeks depending on work load.
Charge per sample. Normal samples (peat, charcoal, wood, textiles, hair, fur, skin), £50. Collagen dates from bone, £70. Shell dates (determined on two separate fractions), £70. Higher charges are made for old dates requiring a large volume of gas and/or a long counting time, but these are unlikely to be incurred by the majority of archaeologists.

BM BRITISH MUSEUM
Address. Dr Richard Burleigh, Research Laboratory, The British Museum, London WC1B 3DG, UK.
Availability. This lab. works exclusively on archaeological dating. Any *bona fide* excavator may apply to have samples dated. Ideally these should be related to one of the lab's existing dating programmes (on which information is available on request), though the lab. is always willing to consider new projects. Advice is taken on new projects from within the British Museum, but the final decision on acceptance is taken by an Advisory Screening Committee, some of whose members are independent specialists from outside the British Museum.
Special interests. The lab. has particular interests in the fields of British and European prehistory in the Palaeolithic and Mesolithic periods and in specific aspects of the Neolithic period and the Bronze Age. The lab. continues to be interested in specific technical problems such as sample pretreatment and comparison of different types of sample material, for example bone and charcoal, which some dating projects simultaneously offer a chance to pursue. More specialised research programmes are pursued in connection with the investigation of natural C-14 variations. The commitment to provide dates for other departments within the British Museum limits to some extent the amount of 'outside' work that can be done.
Turnaround time. About six months.
Charge per sample. No charge is made (though usually no more than ten samples can be accepted from an individual submitter at one time).

GaK GAKUSHUIN UNIVERSITY RADIOCARBON LABORATORY
Address. Gakushuin University Radiocarbon Laboratory, Mejiro, Toshimaku, Tokyo 171, Japan.
Availability. This laboratory runs a completely open dating service. The only proviso is that full information about the sample is given.

Special interests. The lab. has no archaeological specialisations.
Turnaround time. From 8 to 16 weeks depending on work load.
Charge per sample. US $85.00.

GrN GRONINGEN UNIVERSITY
Address. Laboratorium voor Algemene Natuurkunde, Afdeling Isotopen Fysica, Westersingel 34, Groningen, Netherlands.
Availability. In principle the laboratory is completely open, but of course the director reserves the right to be fully informed about the archaeological problems and significance of the samples, and to decide on the proper number of samples and their treatment.
Special interests. The laboratory considers it advisable in the interests of the archaeologist to restrict its activities to areas of its own interest and experience, since it is a laboratory of small capacity. Thus there is a good chance that the lab will be aware of other relevant work in the field and any regional problems that may occur. The area of interest is Europe and the Near and Middle East.
Turnaround time. On average about 6 months, but submitters may mention reasons for priority treatment.
Charge per sample. Dfl. 200.

GX GEOCHRON LABORATORIES
Address. Geochron Laboratories, Krueger Enterprises Inc., 24 Blackstone Street, Cambridge, Mass. 02139, USA.
Availability. Completely open on a commercial basis.
Special interests. The laboratory has considerable experience with bone samples and other less common materials. Facilities for $\delta^{13}C$ measurements.
Turnaround time. Averages around 4–5 weeks from receipt, but can be anything from 1 week to 3 months depending on pressure of other work. The laboratory does, however, recognise legitimate deadline requirements, and where possible, is willing to waive its usual 'first come—first served' rule for valid reasons.
Charge per sample. US $140.00. For $\delta^{13}C$ measurements, requested at the time of submission, an additional US $20.00 is charged. There are no quantity discounts.

HAR HARWELL
Address. Carbon-14/Tritium Measurements Laboratory, Building 10.46, Atomic Energy Research Establishment, Harwell, Oxfordshire OX11 0RA.
Availability. Completely open (notes for the guidance of those wishing to submit samples and application forms available on request).

Special interests. The laboratory is interested in the question of laboratory accuracy of radiocarbon dating. It is seeking higher measurement accuracy through systematic analysis of all aspects of the measurement process, pre-treatment, synthesis, $\delta^{13}C$ corrections and counting. At the same time the laboratory is adamant that the error term associated with its C-14 measurements must realistically express *replicate sample* reproducibility. In collaboration with the British Museum a UK laboratories intercomparison experiment is currently being prepared.
Turnaround time. Normally 3 to 4 months, but a limited service at 6 to 8 weeks can be offered at a nominal extra charge.
Charge per sample. £54 for charcoal, wood and marine shells; £66 for soils and peat; £84 for bones. An excess charge of £10 per sample is levied on samples accepted under the 'express service' scheme provided the result is issued within 8 weeks of the date of reception.

Hv HANOVER
Address. ^{14}C & ^{3}H Laboratorium, Niedersächsisches Landesamt fur Bodenforschung, D 3000 Hannover-Buchholz, Postfach 230153, West Germany.
Availability. Completely open.
Special interests. Besides the general range of interests the laboratory is particularly interested in the application of speleochronology to archaeological dating, for which purpose stalagmitic material is quite suitable. The laboratory is also interested in palaeohydrology and the associated problems of chronology and questions of population movement in arid zones.
Turnaround time. On average 6 to 8 months, with a minimum of 3 months. If it is possible to make advance arrangements with the laboratory free booking of places in the queue is available, thus reducing the turnaround time to the minimum of 3 months.
Charge per sample. No charge when the sample is of general interest to the laboratory. As a commercial arrangement the charge will be about US $180.

I TELEDYNE ISOTOPES
Address. Teledyne Isotopes, 50 Van Buren Avenue, Westwood, New Jersey 07675, USA.
Availability. Completely open on a commercial basis.
Special interests. All types of geochronometric services, including uranium/lead, potassium/argon, rubidium/strontium as well as C-14 dating. Special facilities for dating small C-14 samples.
Turnaround time. Mostly 4 weeks, but varying from 2 to 6 weeks.
Charge per sample. US $160.00. Discount terms for twenty or more samples. For $^{13}C/^{12}C$ isotopic analysis, available on request, an additional charge of US $30.00 is made.

K COPENHAGEN

Address. Radiocarbon Dating Laboratory, National Museum, 10 Ny Vestergade, Copenhagen K, Denmark.
Availability. Samples must be submitted through the National Museum, Copenhagen. They have to pass a screening committee which judges material for its scientific merit and its relevance to Danish archaeology.
Turnaround time. On average 3 to 6 months.
Charge per sample. Dates are not done on a commercial basis.

LJ UNIVERSITY OF CALIFORNIA, SAN DIEGO

Address. Prof. H.E. Suess, Department of Chemistry, University of California, San Diego, La Jolla, California 92037, USA.
Availability. 'Samples have to be of a clearly major scientific interest.'
Special interests. Prof. Suess' laboratory is interested in samples which are directly tree-ring dated.
Turnaround time. 'Two to six months depending on importance of result.'
Charge per sample. None.

Lu LUND

Address. Radiocarbon Dating Laboratory, Department of Quaternary Geology, Tunavägen 29, S-223 63 Lund, Sweden.
Availability. While the main part of the laboratory's capacity is reserved for research work being carried out at the University of Lund, there is capacity for about 30 samples per year from any other submitters.
Special interests. The laboratory can tackle any kind of sample including collagen from bone, horn, antler, tusks and animal teeth.
Turnaround time. About 10 weeks for a series of up to 4 dates, though a week or so longer if the series of samples is larger.
Charge per sample. Swedish kroner 800 (about £84).

NZ INSTITUTE OF NUCLEAR SCIENCES, NEW ZEALAND

Address. Institute of Nuclear Sciences, Private Bag, Lower Hutt, New Zealand.
Availability. Open to archaeologists via the New Zealand Archaeological Association, which acts as a screening committee. Geological samples are submitted via the screening service of the Geological Survey branch of the Department of Scientific and Industrial Research. Proposals from fields other than geology and archaeology should be made direct to the Institute of Nuclear Sciences.
Special interests. Magnetic storage of sample data makes it simple to provide the archaeologist with information on the type, place, activity, etc., of

earlier samples in the categories of his interest. A choice of counting techniques is available for different purposes.
Turnaround time. Depending on the urgency or importance of the sample, waiting time may vary from a few days to two years. On average 12 months.
Charge per sample. For submitters who do not channel their samples through one of the screening organisations a nominal charge may be levied. This charge is NZ $40 per sample for New Zealand submitters, $100 for Commonwealth countries, $150 for the United States of America, $125 for other overseas countries. Exceptions may be made in certain cases, for instance if the laboratory is cooperating on the research project.

P UNIVERSITY OF PENNSYLVANIA
Address. Elizabeth K. Ralph, Department of Physics, DRL/EI, University of Pennsylvania, Philadelphia, Pennsylvania 19174, USA.
Availability. The laboratory accepts samples only with the approval of the appropriate Curator in the University Museum. Since the Museum supports 15 to 20 expeditions per year, there is not much time for dating samples that are not of interest to the Curators.
Special interests. Areas of specialisation are the Near East, Thailand, Mediterranean, South and Central America and the Arctic. The laboratory is also engaged in the dating of tree-ring dated samples.
Turnaround time. Usually about 6 months.
Charge per sample. US $150 per sample; $125 per sample in a series of 8 or more. If samples are of great importance and of interest to the laboratory's specialisations no charge is made. The laboratory strongly prefers to work on series rather than single samples.

Q UNIVERSITY OF CAMBRIDGE
Address. Radiocarbon Dating Laboratory, Subdepartment of Quaternary Research, Botany School, University of Cambridge, Downing Street, Cambridge, UK.
Availability. The laboratory works mostly on projects within the Subdepartment of Quaternary Research and the University of Cambridge, but some samples from outside can be done.
Special interests. None.
Turnaround time. Depends entirely on priorities, research programmes and performance of the instruments.
Charge per sample. None.

SRR SCOTTISH UNIVERSITIES RESEARCH & REACTOR CENTRE
Address. Dr D. D. Harkness, Radiocarbon Laboratory, SURRC, East Kilbride, Glasgow G75 0QU, UK.

Availability. Completely open on a commercial basis.
Special interests. Further development of decontamination procedures, dendrochronological / C-14 correlation in European wood, post-glacial climatic change, mechanisms of soil genesis.
Turnaround time. 3 months, but can be longer for samples which approach the minimum size requirement.
Charge per sample. £70 irrespective of sample size, type or age; price includes mass spectrometric determination of $\delta^{13}C$.

UB QUEEN'S UNIVERSITY, BELFAST
Address. Mr G. W. Pearson or Dr J. R. Pilcher, Palaeoecology Laboratory, Queen's University, Belfast 7, Northern Ireland.
Availability. Samples should either be from Ireland or of specific interest to the research of the laboratory.
Special interests. The laboratory has embarked on an extensive dendrochronologically dated calibration curve and is operating two completely independent dating systems, one capable of achieving precision of ± 35 years and the other capable of $\pm 20-25$ years (dependent on sample size). The higher precision system will not be available for the next 2–3 years but after this period it is hoped to offer it for specially selected samples necessitating this degree of precision.
Turnaround time. On average 6 months.
Charge per sample. £70, including mass spectrometry correction, down to a precision of ± 40 years; for precision better than that £100.

References

Arnold, D. & J. Settgast (1965) *Mitt. Deutsch Arch. Inst. Kairo 20*, 47 (quoted in Berger 1970).

Bauch, J. (1969) Die Bauzeit der Bremer Kogge. *Bremer Hanse-Kogge Monograph 8*, 123-6.

Bauch, J. & D. Eckstein (1970) Dendrochronological dating of oak panels of Dutch 17th-century paintings. *Studies in Conservation 15*, 45-50.

Bauch, J., D. Eckstein & M. Meier-Siem (1972) Dating the wood of panels by dendrochronology. *Nederlands Kunsthistorisch Jaarboek 23*, 485-96.

Bauch, J., W. Liese & D. Eckstein (1967) Über die Altersbestimmung von Eichenholz in Norddeutschland mit Hilfe der Dendrochronologie. *Holz als Roh- und Werkstoff 25*, 285-91.

Baxter, M. S. (1974) Calibration of the radiocarbon timescale. *Nature 249*, 93.

Baxter, M. S. & F. G. Farmer (1973) Radiocarbon: short-term variations. *Earth Planet. Sci. Letts 20*, 295-9.

Baxter, M. S. & A. Walton (1971) Fluctuations of atmospheric carbon-14 concentrations during the past century. *Proc. Roy. Soc. A321*, 105-27.

Becker, B. (1968) The building dates of churches in the Landshut area. *Kunstchronik 21*, 183-7.

Becker, B. (1972) Möglichkeiten für den Aufbau einer absoluten Jahrringschronologie des Postglazials anhand subfossiler Eichen aus Donauschottern. *Berichte der Deutschen Botanischen Gesellschaft 85*, 29-45.

Becker, B. & V. Giertz (1970) Eine über 1 100 jährige mitteleuropäische Tannenchronologie. *Flora 159*, 310-46.

Berger, R. (1970) Ancient Egyptian radiocarbon chronology. *Phil. Trans. Roy. Soc. 269A*, 23.

Berger, R. (1972) Tree-ring calibration of radiocarbon dates. *Proc. 8th Int. Conf. on Radiocarbon Dating*, vol. 1 (eds T. A. Rafter & T. Grant-Taylor) A97-103.

Berger, R. & W. Horn (1970) in *Scientific Methods in Medieval Archaeology* (ed. R. Berger) chs II & III. University of California Press.

Branigan, K. (1970) Wessex and Mycenae: some evidence reviewed. *Wilts. Arch. & Nat. Hist. Mag. 65*, 89.

Bray, J. R. (1972) Cyclic temperature oscillations from 0-20 300 yr BP. *Nature 237*, 277.

Brehme, K. (1951) Jahrringchronologische und klimatologische Untersuchungen am Hochgebirgslärchen des Berchtesgadener Landes. *Z. für Weltforstwirtschaft 14*, 65-80.

Briard, J. (1970) Un tumulus du Bronze Ancien; Kernonen en Plouvorn (Finistère). *L'Anthropologie 74*, 5.

Briggs, C. S. (1973) Double-axe doubts. *Antiquity 47*, 318-20.

Burleigh, R., I. H. Longworth & G. J. Wainright (1972) Relative and absolute dating of four late Neolithic enclosures: an exercise in the interpretation of radiocarbon determinations. *Proc. Prehist. Soc. 38*, 389-407.

Burleigh, R., V. R. Switsur & A. C. Renfrew (1973) The radiocarbon calendar recalibrated too soon? *Antiquity 47*, 309-17.

Butler, J. J. & J. D. van der Waals (1966) Bell beakers and early metalworking in the Netherlands. *Palaeohistoria 12*, 41-139.

Callow, W. J., M. J. Baker & G. I. Hassal (1975) National Physical Laboratory measurements III, in *Proc. 6th Int. Conf. Radiocarbon and Tritium Dating*, Pullman.

Clark, R. M. & C. Renfrew (1972) A statistical approach to the calibration of floating tree-ring chronologies using radiocarbon dates. *Archaeometry 14*, 5-20.

Clark, R. M. & C. Renfrew (1973) Tree-ring calibration of radiocarbon dates and the chronology of ancient Egypt. *Nature 243*, 266-70.

Colquhoun, D. (1971) *Lectures on Biostatistics*, 332. Oxford: Clarendon Press.

Currie, C. R. J. & J. M. Fletcher (1972) Two early cruck houses in north Berkshire identified by radiocarbon. *Medieval Archaeology 16*, 136-42.

Damon, P. E., A. Long & D. C. Grey (1970) Arizona radiocarbon dates for dendrochronologically dated samples, in *Radiocarbon Variations and Absolute Chronology* (ed. I. U. Olsson) 615-18.

Damon, P. E., A. Long & E. I. Wallick (1972) Dendrochronologic calibration of the carbon-14 timescale, in *Proc. 8th Int. Conf. on Radiocarbon Dating*, vol. 1 (eds T. A. Rafter & T. Grant-Taylor) A29-43.

Delorme, A. (1973) Dendrochronologische Untersuchungen an Eichen des südlichen Weser und Leineberglandes. *Forstarchiv. 44*, 205-9.

Eckstein, D. (1974) Tree-ring research in Europe. *Tree-Ring Bull. 32*, 1-18.

Eckstein, D., J. Bauch & W. Liese (1970) Aufbau und Anwendung einer Jahrringchronologie von Eichenholz für die Datierung historische Bauten in Norddeutschland. *Holz-Zentralblatt 96*, 674-68.

Eckstein, D. & W. Liese (1971) Jahrringchronologische Untersuchungen zur Altersbestimmung von Holzbauten der Seidling Haithabu. *Germania 49*, 155-68.

Edwards, I. E. S. (1970) Absolute dating from Egyptian records and comparison with carbon-14 dating. *Phil. Trans. Roy. Soc. 269A*, 11-18.
Engstrand, L. (1965) Excavations at Palaikastro, VI: Appendix. *Annual of the British School at Athens 60*, 314-15.
Ferguson, C. W., B. Huber & H. E. Suess (1966) Determination of the age of Swiss lake-dwellings as an example of dendrochronologically-calibrated radiocarbon dating. *Zeitschrift für Naturforschung 21A*, 1173-7.
Fletcher, J. M. (1968) Crucks in the west Berkshire and Oxford region. *Oxoniensia 33*, 71-88.
Fletcher, J. M. (1970) in *Scientific Methods in Medieval Archaeology*, ch. IV. University of California Press.
Fletcher, J. M. (1974) Tree-ring dates for some panel-paintings in England. *Burlington Magazine 116*, 250-8.
Fletcher, J. M. (1975) Relation of abnormal early wood in oaks to dendrochronology and climatology. *Nature 254*, 506-7.
Fletcher, J. M. & J. F. Hughes (1970) Use of X-rays for density determinations and dendrochronology. *Bull. Faculty of Forestry, Univ. of British Columbia 7*, 41-54.
Fletcher, J. M., M. C. Tapper & F. S. Walker (1974) Dendrochronology—a reference curve for slow grown oaks, AD 1230-1546. *Archaeometry 16*, 31-40.
Fliedner, S. & R. Pohl-Weber (1964) *The Cog of Bremen*. Focke-Museum, Bremen.
Galanopoulos, A. G. (1958) Zur Bestimmung des Alters der Santorin-Kaldera. *Annales géologiques des pays Helléniques 9*, 184-5.
Gimbutas, M. (1965) *Bronze Age Cultures in Central and Eastern Europe*. Mouton.
Girling, M. A. (1974) Evidence from Lincolnshire of the age and intensity of the mid-Devonian temperate episode. *Nature 250*, 270.
Godwin, H. (1962) Half-life of radiocarbon. *Nature 195*, 984.
Harding, A. (1973) Review of *Myrtos* (P. Warren). *Proc. Prehist. Soc. 39*, 476-7.
Harkness, D. D. & R. Burleigh (1974) Possible carbon-14 enrichment in high-altitude wood. *Archaeometry* (in press).
Harkness, D. D. & H. W. Wilson (1972) Some applications in radiocarbon measurement at the Scottish Research and Reactor Centre, in *Proc. 8th Int. Conf. on Radiocarbon Dating* (eds T. A. Rafter & T. Grant-Taylor).
Hayes, W. C., M. B. Rowton & F. Stubbings (1962) in *Cambridge Ancient History 1*, f. 4 (=ch. VI).
Hollstein, E. (1965) Jahrringchronologische Datierung von Eichenhölzern ohne Waldkante. *Bonner Jahrbuch 165*, 12-27.
Hollstein, E. (1968) Dendrochronologische Untersuchungen au den Domen von Trier und Speyer. *Kunstchronik 21*, 168-81.
Hollstein, E. & H. Cüppers (1967) Jahrringchronologien aus vorrömischer und römischer Zeit. *Germania 45*, 70-83.
Hornstein, J. (1964-65) The silver fir timbers from Konstanz and Freiberg minsters. *Alemmanisches Jarhbuch*, 239-89.
Huber, B. (1935) The physiological significance of ring- and diffuse-porosity. *Berichte der Deutschen Botanischen Gesellschaft 53*, 711-19.
Huber, B. (1941) Aufbau einer mitteleuropäischen Jahrringchronologie. *Mitteilung Akad. Deutsch Forstwiss. 1*, 110-25.
Huber, B. (1970) Dendrochronology. *Handbuch der Mikroscopie in der Tecknik 5(1)*, 170-211.
Huber, B. & V. Giertz (1969) Unsere tausendjährige Eichenchronologie. *Sitz. Ber. Österr. Akad. Wiss. 178(1-4)*, 37-42.
Huber, B. & V. Giertz (1970) in *Scientific Methods in Medieval Archaeology* (ed. R. Berger) ch. VIII.
Huber, B. & W. Merz (1963) Jahrringchronologische Synchronisierungen der jungsteinzeitlichen Seidlungen Thayngen-Weier und Burgäschisee-Süd und Südwest. *Germania 41*, 1-9.
Jansen, H. S. (1970) Secular variation of radiocarbon in New Zealand and Australian trees, in *Radiocarbon Variations and Absolute Chronology* (ed. I. U. Olsson).
Jansen, H. S. (1972) The transfer of carbon from solvents to samples, in *Proc. 8th Int. Conf. on Radiocarbon Dating*, vol. 1 (eds T. A. Rafter & T. Grant-Taylor) B63-8.
Johnsen, S. J., W. Dansgaard, H. B. Clausen & C. C. Langway (1972) Oxygen isotope profiles through the Antarctic and Greenland ice sheets. *Nature 235*, 429.
Keeley, L. H. (1974) Technique and methodology in microwear studies: a critical review. *Wld Archaeol. 5(3)*, 323-36.
Kohler, E. L. & E. K. Ralph (1961) C14 dates from sites in the Aegean area. *Am. J. Archaeol. 65*, 357-67.
Kolchin, B. (1962) Dendrochronological methods in archaeology. *Soviet Archaeol. 1*, 95-110.
Kozslowski, T. T. & C. H. Winget (1964) The role of reserves in short growth of red pine. *Am. J. Bot. 51*, 522-9.
Krueger, K. W. & J. M. Trappe (1967) Food reserves and seasonal growth of Douglas fir. *For. Sci. 13*, 192-202.
Lanting, J. N. & J. D. van der Waals (1972) British beakers as seen from the Continent. *Helinium 12*, 20-46.

References

Lawn, B. (1970) University of Pennsylvania radiocarbon dates XIII. *Radiocarbon 12*, 577-89.

Libby, W. F. (1970) Radiocarbon dating. *Phil. Trans. Roy. Soc. 269A*, 1-10.

Luce, J. V. (1969) *The End of Atlantis*. London.

McKerrell, H. (1971) Some aspects of the accuracy of carbon-14 dating. *Scottish Arch. Forum 3*, 73-84.

McKerrell, H. (1972) On the origins of British faience beads and some aspects of the Wessex-Mycenae relationship. *Proc. Prehist. Soc. 38*, 286-301.

MacKie, E. W. *et al.* (1971) Thoughts on radiocarbon dating. *Antiquity 45*, 197-204.

Marinatos, Sp. (1968) *Excavations at Thera, First Preliminary Report (1967 season)*. Athens.

Marinatos, Sp. (1969) Chelidonisma. *Athens Annual of Archaeology 2*, 65-9.

Marinatos, Sp. (1970) *Excavations at Thera, III (1969 season)*. Athens.

Michael, H. N. & E. K. Ralph (1970) Correction factors applied to Egyptian radiocarbon dates from the era before Christ, in *Radiocarbon Variations and Absolute Chronology* (ed. I. U. Olsson) 109-19.

Millar, D. (1963) *Tudor, Stuart and Early Georgian Pictures in the Collection of H.M. the Queen*. London.

Mook, W. G., A. V. Munaut & H. T. Waterbolk (1972) Determination of the age of stratified prehistoric bog settlements, in *Proc. 8th Int. Conf. on Radiocarbon Dating*, vol. 2 (eds T. A. Rafter & T. Grant-Taylor) F27-40.

Muhly, J. D. (1973) The trade routes of the Bronze Age. *Am. Scientist 61*, 404-13.

Munaut, A. V. (1966) Recherches dendrochronologiques sur *Pinus silvestris*. *Agricultura (Louvain) 14*, 193-232 & 361-89.

Müller-Stoll, H. (1951) Vergleichende Untersuchungen über die Abhängigkeit der Jahrringfolge von Holzart, Standort und Klima. *Bibliotheca Botanica 122*, 1-83.

Neiss, W. (1968) Dendrochronology for the building history of the castle and church at Büdingen. *Kunstchronik 21*, 183-7.

Olsson, I. U., ed. (1970) *Radiocarbon Variations and Absolute Chronology*. Proc. 12th Nobel Symposium, Uppsäla, Sweden, 11-15 August 1969. New York: John Wiley, & Stockholm: Almqvist & Wiksell.

Ottaway, B. (1975) Aspects of early copper ore smelting. *Bull. Inst. Historical Metallurgy* (forthcoming).

Ottaway, B. & J. H. Ottaway (1972) The Suess calibration curve and archaeological dating. *Nature 239*, 512-13.

Ottaway, B. & J. H. Ottaway (1974) Irregularities in dendrochronological calibration. *Nature 250*, 407.

Page, D. L. (1970) *The Santorini volcano and the desolation of Minoan Crete*. Supp. paper of the Hellenic Society no. 12. London.

Piggott, S. (1938) The Early Bronze Age in Wessex. *Proc. Prehist. Soc. 4*, 52-106.

Piggott, S. (1973) in *Victoria County History 1(2)*, Wiltshire.

Polge, H. & R. Keller (1969) Xylochronology, a refinement of dendrochronology. *Ann. Sci. For. Paris 26*, 225-56.

Priestley, J. H., L. I. Scott & M. E. Malins (1933) Summer wood production. *Proc. Leeds Phil. Lit. Soc. (Sci. sect.) 2*, 365-75.

Rafter, T. A. & T. Grant-Taylor, eds (1973) *Proceedings of the 8th International Conference on Radiocarbon Dating*, Lower Hutt, New Zealand, 18-25 October 1972, vols 1 & 2. Wellington: Royal Society of New Zealand.

Ralph, E. K., H. N. Michael & M. C. Han (1973) Radiocarbon dates and reality *MASCA Newsl. 9(1)*, August 1973, 1-20. Philadelphia: University Museum.

Rapp, G., S. R. B. Cooke & E. Henrickson (1973) Pumice from Thera (Santorin) identified from a Greek mainland archaeological excavation. *Science 179* (Jan.-Mar.) 471-3.

Renfrew, A. C. (1968) Wessex without Mycenae. *Annual of the British School at Athens 63*, 277-85.

Renfrew, A. C. (1973a) *Before Civilisation*. London: Jonathan Cape.

Renfrew, A. C. (1973b, attrib.) When answers in the soil tell lies. *The Guardian*, 31 October 1973.

Renfrew, A. C. (1973c) Wessex as a social phenomenon. *Antiquity 47*, 221-5.

Säve-Söderbergh, T. & I. U. Olsson (1970) C14 dating and Egyptian chronology, in *Radiocarbon Variations and Absolute Chronology* (ed. I. U. Olsson) 35-53.

Schmidt, B. (1973) Dendrochronologische Untersuchungen an Eichen aus der Kölner Bucht und dem Werre-Weser-Gebeit. *Archaeologischer Korrespondenzblatt 3*, 155-8.

Slåstad, T. (1957) Årringundersøkelser i Gudbrandsdalen. *Medd. Norske Skogforsøksvesen 14(48)*, 571-620.

Smith, A. G., M. G. L. Baillie, J. Hillam, J. R. Pilcher & G. W. Pearson (1972) Dendrochronological work in progress in Belfast: the prospects for an Irish post-glacial tree-ring sequence, in *Proc. 8th Int. Conf. on Radiocarbon Dating*, vol. 1 (eds T. A. Rafter & T. Grant-Taylor) A92-6.

Stuckenrath Jr, R. & B. Lawn (1969) University of Pennsylvania radiocarbon dates XI. *Radiocarbon 11*, 150-62.
Stuiver, M. (1970) Tree-ring, varve and carbon-14 chronologies. *Nature 228*, 454.
Suess, H. E. (1965) Secular variations in the cosmic-ray-produced carbon-14 in the atmosphere. *J. Geophysical Res. 70*, 5937-52.
Suess, H. E. (1970) Bristle-cone pine calibration of the radiocarbon timescale 5200 BC to the present, in *Radiocarbon Variations and Absolute Chronology* (ed. I. U. Olsson) 303-12 and plates I & II.
Switsur, R. (1973) The radiocarbon calendar recalibrated. *Antiquity 47*, 131.
Tauber, H. (1970) The Scandinavian varve chronology and C14 dating, in *Radiocarbon Variations and Absolute Chronology* (ed. I. U. Olsson).
Wade-Martins, P., J. M. Fletcher & R. Switsur (1973) North Elmham. *Current Archaeol. 4(1)*, 22-8.
Warren, P. (1972) *Myrtos. An Early Bronze Age Settlement in Crete*. Thames & Hudson.
Wendland, W. M. & D. L. Donley (1971) Radiocarbon–calendar age relationship. *Earth Planet. Sci. Letts 11*, 135.